The Life of a Leaf

STEVEN VOGEL

The Life of a Leaf

The University of Chicago Press | Chicago and London

The University of Chicago Press, Chicago 60637
The University of Chicago Press, Ltd., London
© 2012 by Steven Vogel
All rights reserved. Published 2012.
Paperback edition 2013

22 21 20 19 18 17   3 4 5 6

ISBN-13: 978-0-226-85939-2 (cloth)
ISBN-13: 978-0-226-10477-5 (paperback)
ISBN-13: 978-0-226-85942-2 (e-book)
DOI: 10.7208/chicago/9780226859422.001.0001

Library of Congress Cataloging-in-Publication Data

Vogel, Steven, 1940–
      The life of a leaf / Steven Vogel.
        p. cm.
      Includes bibliographical references and index.
      ISBN-13: 978-0-226-85939-2 (hardcover :
alkaline paper)
      ISBN-10: 0-226-85939-8 (hardcover : alkaline
paper) 1. Leaves—Growth. 2. Leaves—Physiology.
I. Title
   QK649.V644 2012
   575.5'7—dc23
                              2011037295

To Flora Jane Vogel

Granddaughter with the

most apt of names,

from Grandpa Steve

Contents

HALF A CENTURY of immersion in science has immeasurably enriched my world. I can only describe what it has added with analogies—as how great music sounds when heard in live concert as opposed to the output of a miniature transistor radio. Or as the shift to HD color from fuzzy black-and-white television. Science, I find, is fun to think about, to talk about, to write about, and, most of all, to do. Few activities can measure up to the satisfaction of not just answering some nagging question about the world but at the same time doing something that has never before been done by any human who has ever lived. One even feels empowered when looking at some bit of the immediate world and seeing something that would have been imperceptible without a sense of the scientific.

This book represents an attempt to draw you into that world of science. Not the one of interminable names, arcane procedures, and even more arcane mathematical expressions, but the one that offers satisfying explanations for innumerable aspects of the world around you. Unavoidably, the particular part of that world you'll encounter happens to be my own—"own" not as ownership but as outlook. I'm a biologist who asks about ways in which the appearances and activities of the organisms around us have been determined by their encounters—individually and evolutionarily—with the physical world. That's a world that limits them, one that they can't much alter, but of which they can take all sorts of subtle advantages.

I tend to ask questions close to what might be called intuitive reality, keeping at arm's length big bangs and deep space, atoms and molecules, ultimate causation, and so forth. Thus, relative to more sophisticated science, the price of admission remains low, perhaps close to what it must have been for most of science a few hundred years ago. Really—if I look at the current issue of the oldest of our English-language journals, *Philosophical Transactions of the Royal Society of London*, I can understand only a few of the papers. If I look (online, at my university) at an issue from 250 years ago, I find every paper readable and struggle only to close the volume (or to deselect it). As a result, I find myself concerned with an unusually approachable collection of questions, explanations, and even fodder for investigation. Perhaps you'll be led into that world; better yet, perhaps after all is said and done here, you'll look differently at your immediate world, one that might be at least a little different from mine.

My intent can be put another way. An astronomer or a microscopist might introduce us to an otherwise unseen world. The account here, by contrast, aims to reveal an otherwise unnoticed world. Thus the photographs share a general ordinariness. Almost all are my own, accumulated over quite a few years and places, some opportunistically and some specifically for the present book. They haven't resulted from the patience and artistry of the professional—I'm not long on either virtue. I should admit to a fair bit of manipulation ("Photoshopping" seems to be the current descriptor). And I admit to a fondness for such fix-ups of contrast, background clutter, and so forth—digital, full-color versions of what I once did in the darkroom but reluctantly abandoned when I shifted to color slides.

Every few pages, you'll encounter a suggestion that you might try something at home. If I were teaching a course, it would provide classroom demonstrations, taking shameless advantage of the unusual immediacy of my material. I think all the do-it-yourself suggestions here are safe enough from hazards other than the ire of those with whom one cohabits. If you're lucky, they'll instead be amused.

As to equations, most of us view these as mere formulas into which specific numbers can be put with the expectation that other specific numbers will come out. However, I suggest that the equations given in the footnotes be examined in a different way. Look at what the variables might be to see what matters. Then look at whether each particular one is in the numerator or the denominator to find whether the result will track it or go up when it goes down (and vice versa). And finally, look at the exponents to get an idea of the sensitivity of the result to shifts in the values of the variables.

After all is said, many of the issues here will be left unresolved. For some, this lack of closure can be unsettling, the kind of thing that prompts negative opinions about books and movies. For the working scientist, it generates more positive feelings, ranging from a goad for a specific project to a general sense that yes, science remains an open-ended venture, still fraught with uncertainty and on that account rich with opportunity.

Acknowledgments

I am anything but a proper botanist, even if my degrees say "biology" rather than "zoology," and even if I was (briefly) a member of the Botanical Society of America. I'm not even much of a gardener—my contribution to the family garden consists mainly of compost. So I've drawn on friends, family, colleagues, and prepublication reviewers for information, suggestions, equipment, and extractions of feet from mouth. In particular, I'm grateful for help and tolerance to Marilyn Ball, Anne Benninghoff, Norman Budnitz, Martin Canny, John Close, Mark Denny, David Ellsworth, Ken Glander, Louisa Howard, Peter Klopfer, Andrea Leigh, Dan Livingstone, Margaret McCully, Marty Michener, Adrienne Nicotra, Howard Reisner, Margareta Schmidt-Nielsen, Sanna Sevanto, Miles Silman, Bill Smith, Kathleen Smith, Donald Stone, Jane Vogel, Roger Vogel, Karen Wallace, Dick White, Robert Wilbur, and Claire Williams. Plus our local science book club, where I learn lots about how people read these books—as well as their likes and dislikes. A portion of the book was written while I enjoyed the excellent facilities and surroundings of the Whiteley Center of the Friday Harbor Laboratory at the University of Washington.

WHERE TO START? Maybe before reading further, you should glance out the nearest window. Unless you're stuck in a prison cell or high-rise apartment, you can probably see vegetation, green stuff you ordinarily ignore. It's just life's wallpaper, something that provides a comfortably neutral foreground and softens the starkness and angularity of distant land. Those unassuming bits of vegetation, leaves in particular, provide our present protagonist. I intend to celebrate them, not as poet (Joyce Kilmer comes to mind) or novelist (think of Joseph Conrad), but as scientist. I'll try to convince you that looking at a leaf on a tree from the perspective of a scientist enhances rather than detracts from the aesthetic experience.

I mean to do more than that, however. If the story goes as I intend, you should begin to look with different eyes at your immediate surroundings, seeing not just leaves but yourself and everything around you as reflections of the physical situation here on solar planet number four. Too often we imagine science as a body of facts, growing breakthrough by breakthrough the way a pile of pancakes rises as each new one comes off the pan. At its core, though, science is not the facts but a way of thinking; not a body of knowledge but a way of knowing; a particular and peculiar way of looking at the world. And by "world," we scientists mean more than moons and molecules. We include all the immediate and mundane, things like liquids, lions—and leaves.

As part of this attempt to alter attitudes, I have organized this book in a somewhat eclectic way, so its arrangement asks for a little explanation. Lots of people find numbers and, worse, equations at least off-putting and maybe even indigestible. Other people see them as intrinsic and unavoidable. Since almost all science is inescapably quantitative, we get a severely bowdlerized impression from any account that eschews numbers. While I mean to introduce quantitative arguments as gently as I can, I do mean to include them. Along with numbers (oops), you'll run into equations (horrors). Don't let that bother you, if it's not your métier—the basic text tells the basic story with the sequential linearity necessary for a proper narrative. (Okay, the sequence may be slightly contrived, but after all, this isn't some historical account.) The off-putting details and almost all the quantification have been piled into the footnotes. They're linked by superscripted symbols to the text, so they work rather like hyperlinks on a computer. As a result, you can ignore the formalities in the footnotes without losing the thread of the story. Recognize, though, that graphs and equations provide an economical and effective way of expressing things that torture the tongue. If you read the words and then look at the equation, you'll recognize that they say the same thing. Pretty soon you might start ignoring the words as mere cumbersome redundancy.

I want to encourage the reader to be a player as well, with emphasis on the "play" in *player*. That's the special advantage of asking about matters close to home. So, embedded in the text from time to time you'll find suggestions for things you might do to get a more perceptual feel for what I'm talking about, or to explore beyond what's explicitly mentioned. These "do-it-yourself" interpolations are enclosed in boxes. Again, skip over them if you wish, with no fear of losing the main thread. Finally, to minimize clutter, mention of sources, for both what's here in the text and what's not here but might be of interest, will be relegated to endnotes in the back of the book and indicated by superscripted numbers in the text.

Introducing the Protagonist

The leaf will play a particular—and peculiar—role. It represents a biological everyman, an ordinary and ubiquitous living thing that provides the subject for an exploration of our immediate physical world. We'll look into all (or most, to be honest) of the different physical matters that it has to get right in order to work properly. These are the ordinary phenomena that confront all of us, our domesticated plants and animals, and our mechanical devices. I'd allude to the "cheap physical stuff," except that in my youth that referred to some less savory aspects of human mating behavior. Nonetheless, the word *physical* should be taken more literally than usual. We might look at leaves with biology in mind, asking questions about ecological relationships or about ancestors and lineages. Or we might look at their molecules, at the chemistry of photosynthesis or the genes directing their formation. Here the context will instead be that of more mundane phenomena. Put a bit pretentiously, biological and physical sciences will be inextricably intertwined, as they are in reality as opposed to their dichotomization in high school and college courses.

After all, only in the nineteenth century did scientists adopt the attitude that it wasn't necessary or expected that an investigator be familiar with areas of science in which he or she didn't work. We lost any concern that a well-educated physical scientist might not casually converse with a biological scientist. Curiously, that acceptance of intellectual fragmentation arose at about the same time as the very word *scientist*, originally a replacement for *natural philosopher*, which reflected the earlier fragmentation of philosophy itself. Here I want to revert to the less specialized style of the eighteenth century. In particular, I'll not worry a bit about drawing on not just biology and physics—as currently practiced—but physical chemistry, mechanical engineering, and whatever else puts paint of a pretty color on the canvas.

The best editor with whom I've ever worked (he'll know who he is) advised me to start a book or chapter with a teaser and then move from the specific to the general—not, as in a textbook, from a principle derived up front to examples further along. Teasers, then . . .

- **Intercepting light.** On a summer's day, a sunlit open field feels hot; by contrast you're pleasantly cool in the shade of a forest—even though air moves faster in the field. The difference speaks directly about the effectiveness of light interception by the array of leaves that form the forest canopy. We might take a lesson when designing gazebos, as well as realize how proper eaves and covered porches can improve the comfort of a house in the summer.
- **Not overheating.** Leaves have to absorb sunlight, and they use it inefficiently. So a broad, sunlit leaf in nearly still air can get surprisingly hot. They don't just hang in there, though, but employ a host of devices to keep cool—or at least to keep from getting hotter. Both the devices and the underlying schemes matter to us when we choose cookware, bake at least one kind of pizza, arrange clothing, or pick roofing material.
- **Not being too draggy.** Most of the drag of a tree comes from its leaves. Fluttering things like flags suffer lots of drag—and in the process, as you may notice, fray. But leaves do better by curling and clustering in high winds. We once built large-bladed windmills that permit-

ted some air to pass directly through their blades to reduce their drag when winds got too strong, but we've made little recent use of flexible structures that reconfigure in strong winds or water currents.

- **Getting water up.** Leaves lose lots of water, which the tree must extract from the ground and lift far upward. They use pumps with no moving parts at all. Their scheme pulls water from the top rather than pushing from the bottom. Despite spectacular sucking, they manage to keep air from getting into the system and wrecking everything. We understand their wonderful trick reasonably well, but we've never managed to do much with it in our own technology.

To focus our inquiry, we might put the leaf's basic game in a single (if legalistic) sentence: it uses energy obtained by intercepting sunlight to convert the carbon of the atmosphere's carbon dioxide into larger molecules that can provide material, and, in turn, energy, for growth and reproduction of the plant. The process, as you almost certainly know, goes by the name photosynthesis. We know quite a lot about the basic process and its variations; I mention it here so I can get away with largely ignoring it hereafter. Just don't forget that the criterion for quality—or, we might say, success—for each item that follows boils down to its efficacy in aiding this basic game.

It's a remarkably multifaceted endeavor, this business of doing a leaf's business in a physical world, even if directed at a single end. Assuring access to light, providing mechanical support, coping with heat, deploying from a bud, dealing with wind, getting atmospheric carbon dioxide into the cells, extracting water from soil and raising it upward, deterring herbivores—lots of functions have to be decently done. The diverse devices for doing them can't fail to interact and force compromises, which must be a major reason why the leaves we encounter are so diverse. A list of the physical factors that bear on the leaf's life gets dauntingly formidable: density of plant material, water, and air; viscosity of water and air; mechanical properties such as strength, extensibility, the elastic moduli, and others; thermal capacity, conductivity, and expansion coefficient; surface tension; wind speed; diffusion coefficient; osmotic and hydrostatic pressures—and some others. Every one of these factors bears on your life as well as on that of a leaf—some perhaps less, but most at least as strongly.

Such a complex business doesn't lend itself to a cold plunge into the particulars. It needs some context setting, so here are a few words about each of three nearly independent contexts.

About Science in General

As put a century ago by French mathematician Jules-Henri Poincaré (1854–1912), science isn't about the things but about the relationships among the things. Science tries to see order in the world around us by our best alternative to mutually accepted revelation or mythology. Sometimes that means organized catalogs, things arranged in some arguably natural hierarchy rather than some order-of-convenience-and-convention such as an alphabetical list. More often, and more powerfully explanatory, are rules that apply to a wide variety of overtly disparate and diverse items. The simpler the rule and the wider the range of things it encompasses, the greater its value. The search for predictive and explanatory general rules—that's the crux of our game.

Most often—but certainly not inevitably—our rules involve stepping down in organizational level (or moving up in sophistication, some would say). Thus we explain the motion of the planets, physical phenomena, with mathematical rules; we explain how some substances (visible powders, say) combine with molecular rules; we explain why portholes and aircraft windows have rounded corners with a general explanation of crack initiation and propagation. *Reductionism* describes the scheme, and it has a long history of successes.

No certain sequence defines the reductionist path, though. Should we seek enlightenment by recourse to genetics, to chemistry, to mechanical engineering, to physics, to computational modeling, or to classical mathematics? In a sense, the further "down" the better, with mathematics constituting a kind of grail.* In the end, we're Pythagoreans, engaged in a search for a mathematical order that we believe characterizes the universe. But that ideal provides only the coarsest of guides; were it rigidly prescriptive we'd skip the halfway houses and all become mathematicians. For bet-

---

*I can't resist quoting the title of a paper by a physicist, Eugene Wigner—"The Unreasonable Effectiveness of Mathematics in the Natural Sciences." Beto Cruz, a biomimeticist, puts this "mysterious mathematical harmony" more provocatively: "We hold mathematics as sacred to skirt the appearance of fundamentalism."

ter or worse, traditions take hold—traditions traceable to past successes, to educational inertia, to factors both savory and unsavory.

In biology the dominant tradition has been reduction to molecular chemistry, now including what's come to be called genomics. As an undergraduate, I was advised to take lots of chemistry courses, which advice I dutifully followed since I wanted to become a well-prepared biologist of the next generation. As a graduate student, I happened upon a project to which chemistry had little relevance—I was worrying about the peculiarities of flight in very small insects. Enlightenment came from fluid mechanics, something to which biologists rarely paid much attention. Most of us took a single year of college physics, but the traditional physics course then—and, I think, still—says almost nothing about moving fluids. I knew about viscosity, but I'd heard about it in a course in physical chemistry, not in physics. Step by step, the questions I asked led me into the world of mechanical engineering. A reductionist path, yes, but a different one.

This book intends to make the case for explanation by reduction to physics and mechanical engineering, to this alternative realm of explanation: not to alternative explanations but to explanations of phenomena with which the biologist's classical chemical reductionism just doesn't help. As we'll see, this realm not only explains different phenomena but provides information that makes wonderfully satisfying intuitive sense. Bending, tearing, shadowing, pumping are activities that form parts of our immediate world. When, though, did you last see a molecule? While we assume molecules aren't just polite fictions concocted by chemists, our personal experience doesn't help a lot in thinking about how they behave. Electrons and photons are still worse. In graduate school I roomed for a time with a particle physicist. He ended one attempt to explain the essence of an exciting lecture by admitting, with uncommon candor, that he could think of no explanation, not even an analogy, that wasn't unacceptably misleading. By contrast, I've had the great fortune of working on questions that could be described to just about anyone, from elementary school students to novelists.

About the Biological Big Picture

Evolution by natural selection forms the centerpiece of biology. It's neither physics nor chemistry, so people argue about its position in a reductionist hierarchy. Evolution by natural selection serves here as a background presence, underlying (or haunting) every argument or assertion about how some feature works. It operates this way:

- Reproductive success drives functionally consequential changes and thus much of the design of organisms.
- Reproductive success results from effective functioning of the organism—not just in the mating game but in acquiring resources, growing, and dealing with all aspects of its surroundings.
- Since better functional arrangements lead to greater reproductive success, these arrangements will be favored in the evolutionary sweepstakes.

I like to think of evolution by natural selection as an explanatory principle based on formal logic, an "if, if, and if, then" sequence, because its logical structure conveys the proper note of inevitability. Thus, with no claim of originality . . .

- Observations:
    1. Every organism can produce more than one offspring, so populations, if unrestrained, will increase steadily.
    2. Every organism needs some minimum amount of material from the environment to survive and reproduce.
    3. The material available to a population of organisms is finite in extent, restraining the population's increase.
- Consequence of 1, 2, and 3:
    4. A population in a given area will rise to some maximum size.
- Consequences of 1 and 4:
    5. For a population at this maximum size, more individual organisms will be produced than the environment can support.
    6. Some individuals will not be able to survive and reproduce.
- Further observations:
    7. Individuals within populations vary in ways that affect their success in reproduction.
    8. At least some of this variability is inherited—individuals resemble their parents more than they do more distantly related individuals.
- Consequences of 6 through 8:
    9. Characteristics that increase the number of an individual's surviv-

ing offspring will be more prevalent in the population in the next generation.

Note the repeated use of the word *organism*. Success, or, formally and quantitatively, the concept of fitness, applies almost exclusively to organisms. That's simply because the organism is the reproductive unit. Except in unicellular organisms, one cell can be more fit than another only indirectly—as it might contribute to better organism-level functioning and thus to the organism's reproductive success.

In the present context, a leaf growing at the top of a tree, exposed to full sunlight and thus more productive photosynthetically, can't be more "fit" in this evolutionary sense than a shaded leaf lower down. Leaves may compete for the water and other material moving through the tree's various conduits. But the tree as a whole is the reproductive unit, the generator of acorns or pinecones. The tree may improve its fitness if it makes a shrewd apportionment of resources among its leaves; the leaves contribute like the sentences in a book that's competing for a literary prize. We're individual organisms of a modestly social species, so the picture does no violence to our ordinary sense of personal identity. Bottom line: the hand of natural selection becomes more immediate at the level of the organism than it does for cells, communities, and the like.

The operation of evolution leads to a linguistic conundrum. Evolution has precious little foresight, selecting for what has worked, not what will work. It can only pick for posterity among randomly generated variations. We speak of "design" by nature, although we're quite sure that nature can't design at all—selection has no anticipatory power. Something that might possibly improve reproductive success when further elaborated many generations in the future will not be selected, at least not for that reason alone. Still, *design* describes all too well the exquisitely tuned functional devices of organisms. So, while we use it, we bear in mind its decidedly strange biological meaning. Bad enough as a noun, *design* as a verb misleads so much that I'll try to avoid it altogether.

Circumlocution, though, doesn't solve the basic problem. Organisms simply appear well designed for what they do; in reality they are, in the jargon, "well adapted." If some arrangement seems ill adapted, experience advises us to reexamine our notion of what it does or how it works. Ancestry and other problems afflict natural design: a process that not only lacks foresight but has difficulty with anything but incremental changes will be full of less-than-ideal solutions to its problems. Nonetheless, again, organisms do appear well designed. And so we're led to what's (sometimes disparagingly) called adaptionism. That's the presumption that each feature serves some purpose, that each contributes in its small or indirect way to reproductive success, that none is a mere accident of ancestry or other cause. It can't be exactly correct. Still, what may be in theory a flawed way of thinking turns out in practice to be remarkably effective in generating our working hypotheses. Vision is the purpose of an eye, even if in the most literal sense the eye was not designed to see. In effect, we use *purpose* in the special and restricted sense of contributing to the reproductive success of the parent organism.

The issue has a positive side as well. Some features of organisms contribute to reproductive success, while others turn out to be accidental, trivial, or secondary to some other function. Does it really matter for making offspring whether you're one of those humans who can wiggle your ears or curl your tongue, or whether your earlobes are attached or pendulous? How can we sort features that matter, in reproductive and thus evolutionary terms, from features that don't? Where, in other words, can we see evidence of the operation of the invisible hand of natural selection?

Nature provides us with a tool, one specifically biological and most effective at or near the organism level of organization. It's called convergent evolution. Common features commonly characterize members of groups of organisms that do things in common ways. In a vast number of cases, these common features can't be explained by common ancestry. What's happened is that faced with common challenges, the relevant features of organisms converge. Now, that's a big, bad bugbear when we try to classify creatures based on observable characters. Which truly reflect ancestry and which are mere convergences? The trouble has driven almost all classifiers—systematists—away from using observable characters and toward dependence on more reliably ancestor-reflecting DNA sequence similarities.

By contrast, for anyone studying the functioning of organisms, convergence is a fine thing, a phenomenon that helps us distinguish between what matters and what doesn't. We'll treasure cases of convergence, nature's great gift to the student of organism-level function, unique (to exaggerate only a little) in all of science.[1]

Enough abstractions—we need some specific examples. Creatures are chock full of convergent commonalities . . .

- Along with birds, we mammals have, in a functional sense, two hearts apiece, even if they beat as one. Your left atrium and ventricle pump blood to your body, one organ excepted. Your right atrium and ventricle pump blood to that remaining organ, your lungs. Squid (and other cephalopod mollusks, such as octopus) have a heart that pumps blood to their bodies, again except for one organ. But their arrangement certainly arose separately. A pair of secondary hearts pumps blood to the gills, one heart per gill. What do mammals and birds have in common with cephalopods? All expend energy at high rates by the standards of the larger groups to which they belong, vertebrates and mollusks. That auxiliary heart (or paired hearts) must matter in getting enough blood through their respective oxygenating organs, lungs and gills.[2]
- Treelike plants—columnar, woody things with their photosynthetic structures borne high above the ground—have evolved in quite a few lineages over the past few hundred million years. Trees were a big deal as long ago as the Carboniferous period of the Paleozoic era, and what's left of these large plants provides us with a good part of our coal. They resembled our present ones in height, girth, and so forth, but they differed in a lot of the small stuff. Why trees? Getting leaves a hundred feet closer to the sun takes a lot of material and leaves a plant much more vulnerable to high winds while bringing those leaves no closer to the sun, 90-odd million miles away. We'll get back to the evolutionary rationale for trees further along;[3] the point here is that trees themselves represent a convergent design.
- Leaves living in tropical forests commonly have extended, pointed tips. Leaves of plants that live in dry places are commonly smaller and thicker than leaves from better-watered places. Plants that live in really dry places often have no leaves at all, just thick, photosynthetically active stems. Tree ferns look superficially like palm trees. On tall trees with broad leaves, leaves near the tops are commonly smaller and more deeply indented than leaves near the bottoms. All these commonalities transcend ancestry.

Convergence will be at least a subliminal presence through most of the book. To wax metaphoric, if evolution represents the designing hand of nature, then convergence is the finger pointing to functional significance.

About the Physical Big Picture

Evolution may be prodigiously creative, but it can't tamper with basic physical and chemical rules, nor can it do much with the physical conditions of the surface of our earth. So physical science sets the context, and it needs some prefatory words as well. Since the story here is ultimately about energy, we need an outline of its relevant aspects, one into which we can fit the specifics that follow—the strategy that the tactics aim to support.

First, put aside all thoughts of psychic energy, auras, and similar notions that just wrap themselves in the penumbra of physical reality. Then look up the dictionary definition of *energy*. What you'll find amounts to one or another polite evasion—suggesting, if you take a skeptical view, a bit of a problem. As often put, "Energy is the capacity for doing work." For one thing, that tells us what energy can do rather than what it is. For another, it presumes we know the meaning of *work* in the scientific sense, a sense substantially different from our vernacular verbiage. Put bluntly, any definition that's not misleading lacks easy connection with our intuition.

Why this fixation on energy, not just in the present account but in every discussion about the future of the planet? Mainly, it plays an even more central role in science and technology than does money in economics—even as it provides an analogous accounting medium. Just as money supplies a common scale of value for, say, carrots and cars, energy represents a universal currency for food, metabolic expenditure, and solar radiation. Just as in ordinary transactions money spent by one party equals money received by another, in any physical process, energy lost by energy-expending elements equals energy gained by other elements. Much more strictly than money in any monetarist economic scheme, the total amount of the stuff, whatever energy really is, remains unchanged. There's no such thing as a free lunch.

The notion of energy permits us to assert a conservation law, and for the scientist, that's ample justification for making a big deal of it—as we do with conservation of mass, conservation of momentum, and conservation of some lesser-known variables. Put formally, in any process the total energy remains unchanged. And given a formal name, it's the first law of thermodynamics. Energy, of course, comes in quite a number of forms: mechanical energy, which is energy by virtue of mass in motion; chemical potential energy, energy that can be released (really that can be shifted to something else) in a chemical reaction; gravitational potential energy, what comes out when something goes downhill; thermal energy, energy that shifts if a hot body is in contact with a cooler one; and others.

If I walk on an inclined treadmill, I expend energy doing mechanical work on the treadmill (and its motor consumes just a little less electricity when I'm on it), and I release some extra body heat into the room. Together these add up to the little (too little!) extra food that I can eat without increasing the stored energy of my body. All of this should sound exceedingly ordinary and familiar. Ordinary even if we're assailed with promises of better outcomes from purveyors of one or another diet plan—who truly live off the fat of the land. And so familiar you're probably wondering why I'm dwelling on it.

For present purposes, the first law provides a stepping-stone to the critical but more subtle second law of thermodynamics. This one, rather than being a conservation law of energy concerned with total energy content, defines the direction in which energy flows. It lends itself less well to a succinct statement in words, but a general description agrees well enough with everyday experience to sit easy on the intuition. Perhaps it's best introduced by asking the most ordinary of questions: in what direction do basic processes go if left to their own devices? The second law asserts that the natural, spontaneous direction for processes in our universe is (1) from higher to lower potential, as when water flows over a dam or heat flows from a hot body to a cooler one, and (2) from greater to lesser order—or lesser to greater disorder, as in its formal statement.[4] The title of Chinua Achebe's novel, *Things Fall Apart*, alludes to this second kind of spontaneous process.

Back when I taught introductory biology, I'd illustrate the second law by dropping a dish from the lectern onto the floor, noting that this was not an analogy or metaphor but a proper example—inasmuch as we were inescapably embedded in the universe. Without fail, the plate reduced the distance between itself and the center of the earth, and (assured by a high enough drop) became notably less ordered in the process. No one ever asked that I drop the accumulated pieces to see if they would spontaneously reassemble—none of us are that ignorant of the rules of reality.

But the demonstration, whatever its virtue in focusing the attention of an early-morning lecture class, did miss a key aspect of the second law. Why *thermo*-dynamics? That's because heat plays a special role, one different from other forms of energy. Heat represents the average rate of random movement of molecules, wandering in liquids and gases, vibrating in solids. Random movement—that's the very essence of disorder, a word just used. The second law asserts that in any process, thermal energy increases at the expense of other forms of energy. Your electric space heater is 100 percent efficient, with all of its 1500 watts (or whatever its rating) shifting from electrical to thermal forms.* But no motor can convert all the input of fuel or electricity into mechanical work—it must leak some heat as it operates.†

While the second law may resist strict proof, no clear violation has ever been observed. James Clerk Maxwell (1831–1879), a British physicist, contrived an instructive example of an apparent violation a century and a half ago, a fine mind's-eye picture. Imagine, as in figure 1.1, two water-filled chambers separated by a wall.

If one contains hot water and the other contains cold water, the system will gradually equilibrate, leaving lukewarm water in both—a less-ordered situation. Thereafter, the pair will never move back to their original hot-plus-cold condition. Unless . . . What if some creature, familiar to physicists as "Maxwell's demon," now operates a tiny gate between the cham-

---

* An advertisement for quite an expensive space heater has been appearing in many newspapers and magazines as I write this. As a result of great breakthroughs, it can produce "an amazing 5119 BTU's per hour of heat." A quick lookup of the relevant conversion figure (obtained from *Wikipedia* or other source) tells me that that output is exactly (too exactly!) what we get when plugging in any electrical device that draws 1500 watts—such as a resistive heater that costs less than a tenth as much.

† The first law of thermodynamics can be encapsulated as "You can't win." The second law, treated similarly, says, "You can't even break even." There's a third law, one we don't need here, that amounts to saying, "And you can't get out of the game." Another summary of the second law, recognizing the special role of heat, asserts that "sooner or later, everything goes to hell."

Incidentally, the second law is one of the few physical rules that require time to run in one direction rather than the other—it describes how processes must proceed.

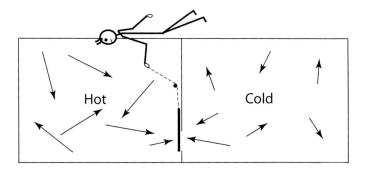

FIGURE 1.1. Maxwell's demon oversees a gate between two compartments in which water molecules move around—some faster, some slower. The demon opens the gate to let any relatively slowly moving molecule move to the right, or let any relatively rapidly moving molecules move to the left. As a result, the hot water gets ever hotter and the cold gets ever colder.

bers? What we call temperature represents the average speed at which molecules move in a substance. So on average, hotter water has faster molecules. Say the demon lets slower molecules move right and lets faster molecules move left. Eventually, one chamber will hold hot water, the other cold, as at the start. But several persuasive arguments disallow this violation of the second law. In particular, the demon has to learn something about molecular speeds, and getting information takes energy. So the demon's own energetic demand cancels out anything that might be gained.

With an input of energy, order can be increased. In this example, we simply put a heating coil in one chamber and a cooling coil in the other, putting in energy on one side and draining it out on the other—of course at some operating expense. Voilà—order, in this case realized by coupling one energy flow, the one running the coils, with another, the net movement of heat from one chamber to the other. You might retort that this is a pretty poor grade of order. If that's the way you feel, here's a little more impressive case, one, moreover, that could not be closer to home. Heating a pan of water on a stove makes heat flow from the water at the bottom, where it's put in, to water at the top, from which some of it moves out to the room. Energy is flowing, and the water molecules develop something nicer than mere random movement—distinct convective patterns such as what are called Bénard cells. Figure 1.2 shows such cells, both as an idealized diagram and as photos taken of my approximations in our kitchen.

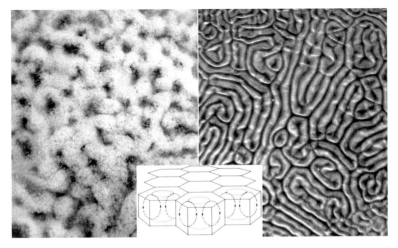

FIGURE 1.2. Heat flow can generate order. The drawing shows the textbook, idealized hexagonal array of Bénard cells, while the photographs show what I produced on the kitchen stove: the one on the left, using miso soup (Kikkoman Tofu Miso); the right-hand one, pearlescent soap (Bath & Body Works Mango Mandarin Creamy Body Wash). (The photographs have had their contrast boosted.)

Energy flow has generated order![5]

Do it yourself: All you need is a pan with a dark inside bottom, a source of heat, and a small container of so-called pearlescent or opalescent soap. Mix a teaspoon or so of the soap into about an eighth inch of water in the pan, and set the pan over a very gentle heat source—perhaps use a larger pan with hot water in it, or support the pan so that its edges are well above a burner (on low) of a stove. Wait a while for residual currents to damp out before turning on the heat. You might see what changing the conditions does to the patterns that develop. Incidentally, dilute pearlescent soap also produces lovely vortices behind moving objects.

Note: Pearlescent liquid soaps aren't as common as when they were still a novelty. A grooming shampoo for dogs, Mycodex, ought to work as well as the body wash my wife generously provided. On ingredient lists look for glycol stearate, the common pearlescent agent. Sometimes the pearlescence can be seen through the glass or plastic jar.

You must be wondering what all this has to do with leaves. To be cagey no longer, it puts us in touch with the fundamental reason why leaves matter. One of the worst misrepresentations of science by creationists is their claim that the second law's prediction of perpetual deterioration of order shows the need for divine intervention if life is to appear and persist. No—if energy flows, order can increase, as in that pan of soapy water or miso soup. What's minimally needed to generate order are three items: (1) a source of energy and (2) a sink for energy, with the latter at a lower potential (cooler or lower down) than the source, and (3) some coupling system to draw on this energy flow.

For our earth, the sun provides the source, and the sink is outer space or, in immediate terms, the cold sky. What's the coupling system? One system exceeds in importance by some vast factor all others put together. It's photosynthesis, as done by green plants, algae, and some kinds of bacteria. Without photosynthesis (or some substitute), nothing like the present kind of complex, highly ordered life could exist. Leaves are *really*, *really* important.

Now take a more respectful look at some leaves, even if only the bay leaves on your spice shelf or the green stuff in your salad.

IT'S NO ACCIDENT that we raise our major food crops in full sunlight, planting only a few specialized edibles, such as coffee beans and peppercorns, in the shade. The basic game for a leaf, again, consists of capturing sunlight and storing it, not as heat to be reradiated to the far reaches of outer space, but as the energy necessary to make medium-sized molecules out of small ones. Those medium-sized ones can then be broken down, channeling the energy into other thermodynamically unfavorable activities such as building still larger molecules. These latter activities may be done by the plant itself, or it may supply power to the hangers-on—nongreen epiphytic plants, soil microorganisms breaking down dead plant material, herbivorous animals directly, and carnivorous animals indirectly but quite as surely.

The key to these coupled processes was provided in the 1770s by Joseph Priestley—scientist, radically liberal cleric, and political refugee. We hear too little about Priestley, at one time member of the "lunar men" of Birmingham, England, and later a Pennsylvanian who corresponded extensively with both John Adams and Thomas Jefferson. Posterity poses problems for anyone who becomes an anachronism late in life. Priestley stuck with phlogiston, the ostensible product of burning, long after Antoine Lavoisier persuaded the scientific community that burning added something— oxygen, which Priestley himself had first isolated—rather than producing any such thing. But about what plants did to air, Priestley got it right. A mouse in a closed container became distressed and died after a time, and subsequent mice died immediately in that air. But mice lived a lot longer if plants had grown in the container. In our terms, the plants produced oxygen, which the mice used up.* Plant growth required that the container be illuminated, but the particular contribution of sunlight remained obscure, so obscure that

---

*"This observation led me to conclude that plants, instead of affecting the air in the same manner with animal respiration, reverse the effects of breathing, and tend to keep the atmosphere sweet and whole-

Seeking Illumination

2

it drew no remark. How few of us gratuitously conduct our experiments in the dark?

What Light Should a Leaf Absorb?

We're vaguely aware that light varies in more than simple intensity. Things take on slightly different colors with different illumination sources. At one time we bought color photographic film color-balanced for either sunlight or incandescent lightbulbs, depending on our intent. Using the wrong film meant distorted colors in the photograph. With digital cameras we can select from a menu of possible color balances.

The character of sunlight, filtered through the atmosphere, depends on the sun's angle—the time of day—as well as what pollutants it encounters. The light available in shade depends, in addition, on the character of the various surfaces whose reflected and scattered light illuminates the shade. We can buy so-called Grow Lights (or "Gro-lights"), matched to sunlight or to the tastes of green plants by their manufacturers. (Both the claimed match and their efficacy, though, have been debated.)

What's behind this business of matching, this vague notion of light quality? Mainly what we ordinarily describe as wavelength. That notion needs a little explaining. Light is a form of electromagnetic radiation like radio waves, but unlike sound waves or the surface waves on water. It shares (confusingly and counterintuitively) some properties of particles (called photons) and some properties of propagating, traveling waves. By *waves* we mean oscillations or vibrations at regular and particular rates—or frequencies. Something that oscillates at the same time as it travels steadily

---

some when it is become noxious, in consequence of animals living and breathing, or dying and putrefying in it."

Priestley recognized the biospheric implication; Benjamin Franklin, then in London and ever alert to the practical, wrote, "I hope this will give some check to the rage of destroying trees that grow near houses, which has accompanied our late improvements in gardening, from an opinion of their being unwholesome. I am certain, from long observation, that there is nothing unhealthy about the air of woods; for we Americans have everywhere in our country habitations in the midst of woods, and no people on earth enjoy better health or are more prolific" (Priestley and Hey [(1772], pp. 193, 199–200, these last quoting Franklin). Johnson (2008) views the work in context; the standard biography of Priestley is Schofield's two-volume work (1997, 2004).

can be described as having a wavelength. For instance, if you mark a long blackboard (or whiteboard) by moving a piece of chalk (or marker) up and down while walking alongside it, you draw a wave whose wavelength, peak to peak, is given by your walking speed divided by the frequency of your up-plus-down arm movements. If you walk at 4 feet per second while making 2 up-and-down movements each second, you make waves of a wavelength of ½, or 2 feet—try it if you need experiential persuasion.

Something that vibrates more rapidly takes more energy. Similarly, more energy can be extracted from something that's vibrating more rapidly—like light (whatever its actual nature). The higher the oscillation frequency of the light, the more energetic are its photons. And, as with the line on the blackboard, the higher the frequency, the shorter the wavelength. Light travels, whether through the air or in a vacuum, at 300 million meters per second (670 million miles per hour). So frequency times wavelength has to equal 300,000,000 if we stick to meters and seconds (as the relevant authorities urge).* Midrange visible light, 500 nanometers (500 US billionths of a meter, or 500×10^{-9} meters) in wavelength, vibrates 600 trillion, or 6×10^{14}, times per second. As every photosynthetic leaf knows well, it's pretty energetic.

By comparison, the carrier signal of my favorite NPR FM radio station vibrates 92 ½ million, or 92.5×10^{6}, times per second (92.5 on the dial, assuming the millionfold multiplier). Our microwave oven's power unit vibrates 2.45 billion, or 2.4×10^{9}, times per second. The "micro" in *microwave* refers to its wavelength, short relative to radio or television waves, but far,

---

*The formula, in formal terms:

$f\lambda = c,$

where f is vibration frequency, λ (lowercase lambda) is wavelength, and c is the speed of light.

Science generally works with SI (for Système Internationale) units. The basic ones are meters (for measuring length), seconds (time), and kilograms (mass). Other units emerge as combinations of these. Thus acceleration has units of meters divided by the square of the number of seconds. Derivative units have then been defined so that as many basic formulas as possible have no funny multipliers to remember—as in the one above. For instance, a formula, force = mass times acceleration, defines a unit of force, the newton. In SI units, the newton then equals one kilogram-meter per second squared.

For the formula above, frequency is given in hertz, or 1/seconds, wavelength in meters, and the speed of light in meters per second.

far longer than light waves. The first common brand was appropriately called a Radarange—radar uses wavelengths similar to the 12-centimeter ones of microwave ovens.*

Visible light shades smoothly from shorter, more energetic wavelengths of about 400 nanometers, which our eyes perceive as violet, through blue, green, yellow, and orange to red, at about 700 nanometers. Shorter, invisible, but more energetic wavelengths travel as ultraviolet—beyond violet—and tend to bust up molecules. They constitute "ionizing radiation," and you want to keep your exposure to them as low as possible. Longer, less energetic wavelengths pass as infrared—under or after red. They penetrate biological material somewhat better, but they do little or no damage in the process except to the extent that absorbing them heats the material. Sunlight that has come through our atmosphere is a mix of all three kinds, as in figure 2.1. (The vernacular gets it right, dis-

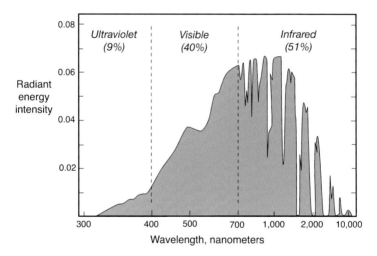

FIGURE 2.1. The spectrum of solar energy arriving at the earth's surface. The odd-looking scale on the horizontal axis causes equal areas of the graph to correspond to equal amounts of energy. (So-called wave numbers would scale linearly on such a graph.) So, you can get an idea of how much energy comes in the various spectral regions—as given in the percentages at the top.

* At one time, perhaps still, radar stations in the Canadian far north watched for the Soviet bombers we were assured posed a threat. In the 1950s, technicians working at them would climb onto the antennas on cold days and get pleasantly warm right through heavy clothing. Development of odd heat-related pathologies such as eye cataracts put an end to the indulgence.

tinguishing between sunburn, from too much ultraviolet, and heatstroke, from too much visible and infrared.)

The way the energy of radiation varies with wavelength (or frequency) gets us back to the comments on thermodynamics in chapter 1. Light of more energetic short wavelengths can be used to generate light of less energetic longer wavelengths with no great difficulty. Fluorescent lightbulbs produce ultraviolet light inside themselves, which the coatings of the bulbs convert to less energetic visible light. Toothpastes and laundry detergents commonly down-convert to visible light whatever ultraviolet light strikes tooth or tablecloth, so the advertising slogan "whiter than white" does have some truth to it. Conversely, shortening the wavelengths of radiation requires energy, so night-viewing devices that illuminate a scene with invisible infrared need an energy input to up-convert that infrared to something visible in goggles or on a screen.

Now, back to leaves. Photosynthesis requires solar energy in the form of light. "In the form of light" sounds redundant but is not. Usable wavelengths—light at the right energy level to be fed into the process—correspond fairly closely to light as perceived by the human visual system, 400 to 700 nanometers. Since some other animals' visual systems have other ranges of spectral visibility, that must be partly coincidental. The visual spectrum of insects, for instance, runs toward shorter wavelengths. They see near ultraviolet, which we don't, but they don't see our familiar red. Flower colors, which matter to many insects, look different to insects and to us, and insect-pollinated flowers are rarely red—red flowers more often attract hummingbirds.

Only about half the energy of solar radiation arrives at the earth's surface in the form of visible (to humans) and photosynthetically active (to green plants) wavelengths. As the percentages in the figure tell us, most of the rest comes in as near infrared, with wavelengths just a bit longer than those we can see. It's not energetic enough to do much in the way of proper chemistry. Most common materials absorb it, warming in the process. Heat lamps produce a minimum of visible light and a lot of infrared, so we can warm such things as outdoor diners, filled dinner plates awaiting the waiter, and incubating eggs without too much visual dazzle. By contrast, leaves respond in an unusual way to all this solar near infrared—they're remarkably nonabsorptive. This light doesn't go through as if they were transparent, but for the most part it gets reflected and scattered at the leaf surface. The spectral transition for a leaf is sharp, with absorption abruptly ending just beyond wavelengths that have sufficient energy for photosynthesis.

That phenomenon matters enough to hint at adaptation and natural selection. Again, absorbing radiation warms a body, which, for instance, is how microwave ovens work. That warming presents a problem for leaves. If fully absorbed, direct sunlight would heat leaves quite a lot hotter than most can tolerate—the overhead sun can deliver a whopping 1000 watts per square meter, almost 100 watts per square foot. A broad leaf in direct sunlight and windless air can get as much as 20°C (36°F) above the local air temperature. When the air around it reaches the mid-30s (mid-90s F), the leaf tickles temperatures at which proteins start changing shape in undesirable ways. An animal can seek shade or use its bulk to slow the rate of its body heating—camels, most notably, do this only as a last resort, confident that cool nights follow warm days—but a leaf can't hide, and with minimal volume for its surface, it heats up quickly. We'll get back to temperature and heat dissipation in a few chapters. Here we focus on that abrupt shift from light absorption to nonabsorption at a wavelength of around 700 nanometers.

It has a practical consequence for human technology. We can buy monochromatic (black and white) photographic film that's sensitive out to about 900 nanometers, well beyond our visual limit of 700 or so; it's called infrared film. Ordinarily, it's used with a red filter on the camera to limit its exposure to visually perceived wavelengths—the red filter cuts out most light of wavelengths shorter than those of red. Infrared film sees through bluish haze somewhat better than ordinary film, and it has some other special-effect uses. Blue sky, for example, comes out dark on a positive print—it's white or nonexposed on a negative, since it doesn't emit much red or near infrared. Clouds, then, contrast strikingly. Water absorbs near infrared, so a light sandy beach contacts strikingly black water. Human flesh and foliage reflect it, so they come out lighter than we'd normally see on positive prints, as in figure 2.2. (A color infrared film also exists; on it skin is light green, skies are blue-green, and foliage is magenta. Not exactly a pretty picture.)

Aha—foliage. Back to the central player here. Trying to keep cool, foliage rejects infrared. Absorbing photosynthetically useless energy would be all cost and no benefit. In a photograph made using infrared film, a

Visible light   *Enhanced infrared*

FIGURE 2.2. Three views of our fig tree against the sky. The color image in the center was converted to the black-and-white one on the left. A camera loaded with infrared-sensitive film (Ilford SFX 200) and equipped with a red filter (Kodak Wratten A) that suppressed short-wavelength light produced the image on the right. I've adjusted the pictures so that the sky is the same shade of gray in the two black-and-white views.

house or a gravestone jumps out of almost pure-white shrubbery. More consequentially, green foliage looks quite different in the near infrared from other, ordinary, green materials—near white versus near black. The military values that distinction between real foliage and camouflage, perhaps the impetus for development of infrared film.

Do it yourself: Try the trick using a roll or two of infrared film, which is still available, at least for 35-millimeter cameras. Infrared film "sees" a region in the visible realm as well, and this needs to be eliminated by putting a red filter in front of the camera lens. Proper focus will be just a little different, since lenses bend light of different wavelengths by different amounts. That won't matter a lot for work in bright sunlight, but it's the reason digital cameras commonly incorporate an infrared-rejecting filter and can't easily be pressed into the present service—even though the sensors themselves respond to infrared. While it's possible to get the filter out, it's not something casually undertaken.

So leaves reject both most of the ultraviolet that might do chemical damage (ultraviolation?) and most of the near infrared that might do thermal damage. At the same time, they manage to funnel visible light of a wide range of wavelengths into their photosynthetic gear. Chlorophyll absorbs visible light in two regions, a blue band at around 430 nanometers and a red band around 680, with only minor bands of absorptivity in between. So by itself it's not very sensitive to most of the light coming through the atmosphere. Leaves, though, contain accessory photosensitive chemicals, mainly carotenoids, which absorb in the blue, yellow, and, to some extent, green. Recall that light energy—photons—can pass from shorter to longer wavelengths. That permits linking the accessory pigments to chlorophyll, ultimately because photosynthesis doesn't really need light with anything shorter than a 680-nanometer wavelength.*

Somewhat oddly, leaves absorb relatively little at wavelengths in the green range, around 510 nanometers, even though they can feed both shorter- and longer-wavelength sunlight into their photosynthetic machinery. That paradoxical rejection of the green makes our leafy world pleasantly green rather than monochromatically black. It's not an obviously functional device, so either some chemical difficulty presents itself or—what's hard to reject—it may simply be an evolutionary accident. For leaves, green is anything but the color of money and certainly doesn't signal "go."

We've one more transaction in this business of radiation, wavelength, and energy content. What determines the wavelength of radiation that a body will emit? As it happens, the higher the temperature of the body, the shorter will be its predominant emission wavelengths. The surface of the sun has a temperature of about 5500°C (5800 K),† which puts its

---

*Nobel puts some numbers on the photochemistry. Light at 680 nanometers corresponds to photons containing 176 kilojoules of energy (per mole, the relevant unit of quantity, which we'll leave undefined here). It takes eight photons to capture a molecule of carbon dioxide, or 1408 kilojoules per mole. Of this, 479 gets stored by that chemical fixation, so the process is 479 ÷ 1408, or 34%, efficient—remarkably good as these things go. But the overall process, beginning with sunlight and ending with sugar, doesn't come close to this efficiency.

†K stands for kelvin, after Lord Kelvin, William Thomson (1824–1907). It's the temperature scale used in physics (with no degree sign, °), because it has a nonarbitrary lower limit, the temperature at which molecular motion ceases ("absolute zero"). The

greatest emission, the brightest light, in the yellow part of the visible spectrum.*

I alluded earlier to reradiation to the sky and outer space. A leaf at 30°C (86°F, or 303 K) or a human with that skin temperature will reradiate at about 10,000 nanometers, or 10 micrometers—a lot longer wavelength than the near infrared we've been talking about. How much radiation is emitted in that region depends on something called the emissivity, with a scale running from 0 to 100 percent. Most materials are quite emissive in this range, with values around 90 percent. Foliage emits with particular effectiveness, with values between 95 and 98 percent. That can offset some of the heating effect of direct sunlight, especially if the sky is clear. A clear sky is not quite the same as outer space, which returns no radiation at all. Clear sky radiates at an effective temperature around –53°C (–63°F; 220 K). But a cloudy sky is worse, with an effective temperature around +7°C (45°F; 280 K).†

Since I became bald, I've been sensitive to that difference, at least when standing in a windless place at night—my scalp tingles a bit when the sky above is clear. It's a more serious matter for leaves, particularly ones close to the ground, where air movement can be minimal. On clear nights, frost

---

units are just the same size as Celsius units, so one just adds 273.16 to the Celsius temperature to get the kelvin temperature. US readers need first to switch from Fahrenheit to Celsius, for which $C = \frac{5}{9}(F - 32°)$.

We much prefer to work with scales that start from zero, for quite a simple reason. You may be able to say that 30°C is 10°C hotter than 20°C, but the statement that 30° is 50 percent hotter than 20° is just a misleading artifact of the particular scale. (For instance, in Fahrenheit units, that 50 percent would become 26.5 percent.) 30 K (brrrr!), though, truly is 50 percent warmer than 20 K (brrrrrrr!), and it would be 50 percent even if Fahrenheit-sized degree increments were used instead of Celsius-sized increments.

*The relationship between temperature, T, and wavelength, λ (lambda, again), is simple, especially if using the kelvin temperature scale. It's called Wien's law after W. C. W. O. F. F. (!) Wien (1864–1928), a German physicist:

$$\lambda_{max} = \frac{0.0029}{T}$$

† The net exchange of radiation is given by the Stefan-Boltzmann equation,

$$q = \sigma S(\varepsilon_2 T_2^4 - \varepsilon_1 T_1^4).$$

q is the rate of energy transfer (joules per second, or watts, in SI), σ (lowercase sigma) is a constant, 5.67×10^{-8}, S is the exposed area, ε's (lowercase epsilons) are emissivities, and T's the temperatures of the two objects.

can form on their surfaces, even if the overall air temperature doesn't quite reach the freezing point (fig. 2.3). Worse, ice crystals can develop inside them. Less worrisome or even beneficial (and fortunately more common) is dew, condensation of moisture that occurs as the leaves' surface temperature drops below the dew point of the surrounding air.

FIGURE 2.3. Frost granules on a flower and some leaves in a botanical garden in Dunedin, New Zealand. The cold snap that generated these occurred in July 2003, disconcerting to us Northern Hemispherics as well as, it appeared, the local ducks, whose swimming pond froze over.

We too commonly ignore radiative exchange among objects at less than light-emitting temperatures, but it's always happening. I recently bought an inexpensive remote infrared thermometer, a point-and-shoot handheld device. Measurements made around my yard during calms left no doubt that surface temperatures, and not just those of leaves, could be quite different from air temperatures. On one cool winter morning just after dawn, the air was –2°C (29°F), while surfaces ran between –9° and –13°C (16° and 9°F). Magnolia leaves, sheltered and low to the ground but exposed to the sky, were the coldest items I found. On another morning in another place, I recorded magnolia leaf temperatures as low as –20°C (–3.5°F), with an air temperature of –8°C (18°F).

This radiation to a clear, cold sky might underlie all those "cold moon" allusions and metaphors. Only with a clear sky can we see the moon, so it seems to radiate cold in the same way the sun radiates heat. Perhaps only cultures for whom all nights are clear (arid lands, mainly) will lack such allusions and metaphors. Sun and moon have traditionally been seen as opposites, hot and cold, with the sex of the respective deities associated most often with the opposite of that of the other. Thus "Chanting faint hymns to the cold, fruitless moon," in Shakespeare's *A Midsummer Night's Dream*.

Do it yourself: These point-and-shoot infrared thermometers are nifty gadgets, well worth having around the house to check insulation effectiveness, cooking surface temperatures, and lots of other interesting thermal items. Mine is a Raytek MT6 and cost about $50US; I doubt whether brand and model carry much significance. So you, too, can look at leaf temperatures in a marvelously quick and nonintrusive manner.

My mother would gauge the coldness of a winter morning by looking out the kitchen window at our rhododendrons. On cold nights, and circa-zero nights (0°F; around –20°C) were all too common in the Hudson Valley, their leaves dropped down to a near-vertical orientation and rolled up into cylinders. The colder the night, the tighter the roll, which sometimes approached the diameter of a pencil—as in figure 2.4.

Who needed a thermometer? The phenomenon isn't just family lore—the tightness of the curls has been shown to track (inversely) local tem-

FIGURE 2.4. A group of rhododendron leaves on the Duke University campus, around mid-day (*left*) and then at a cold dawn (18°F, –8°C; *right*). Exploring with the infrared gun during the same dawn, I measured leaf temperatures as low as –3.5°F (–20°C), the extremes on the leaves of a nearby magnolia. Pop a few unrolled leaves in the food freezer, and they rapidly roll—and even more rapidly unroll when removed.

perature. I see the same thing, if in a less extreme form, if I glance at the *Aucuba* in my present yard. Erik Tallak Nilsen pointed out that the most probable function of curling is minimizing exposure to the sky. That, as well as the vertical repositioning, would keep leaf temperatures from dropping quite as far below the temperature of the surrounding air as would happen otherwise. He noted that rolling up differed from wilting downward, both in mechanism and in consequence; that different species of the genus *Rhododendron* varied in their propensity to curl; and that the cold-tolerant ones did it best.[1]

Something even neater may be going on, something that (as far as I know) still awaits investigation. The air within those rolled-up cylinders will be cooled by its proximity to the leaf surface. So it will be cooler than the air outside the leaf, which will be mixed with the rest of the local atmosphere. As a result, it will be a little denser than the rest of the air and will therefore sink, drawing more ambient air in at the top of the leaf cylinder, as in figure 2.5. This flow—we'll go into more detail about convec-

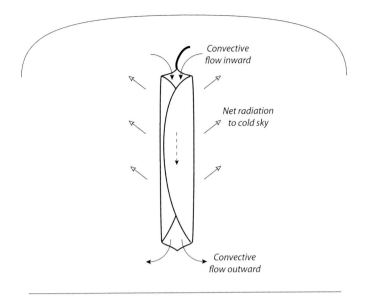

Convective
flow inward

Net radiation
to cold sky

Convective
flow outward

FIGURE 2.5. A rolled leaf as an antichimney. Radiation from the leaf to a clear, cold sky drives the leaf temperature below that of the surrounding air, which increases the density of air within the rolled leaf and generates a downward flow.

tive flows in chapter 8—will help keep the temperature of the leaf near that of the air. Moreover, it will help warm—if slightly—even the unfortunate part of the leaf surface that, despite rolling up, remains exposed to the sky. What I'm suggesting, then, is that these cylinders are antichimneys. Locally hotter air rises, locally colder air sinks.

Low-speed airflows through and around organisms have provided me with research material for over half a century. So I force myself to realize that my enamorment with hypothetical antichimneys may just be the normal bias of a fan of flow—if your tool is a hammer, the problems look like nails. My ever-percipient friend Marty Michener suggests, to get something else on the table, that curling might just be a trick that allows the leaves to shed a coat of ice. That can happen when a cold front comes through, and the temperature drops following a freezing rain. I can identify with that, since a severe, windless ice storm in 2002 dropped two trees on our house. More on ice in chapter 10.

Sun versus Shade

Once again, the basic game for a leaf consists of using solar energy—sunlight—to synthesize large molecules from small ones. Some leaves do best in full sunlight; others do better at the lower light levels of shady places. We can have too much of a good thing, and full sunlight isn't automatically tolerable—as those of us with fair skins know too well. Still, those leaves that live in the shade never achieve the levels of chemical activity of the ones in direct sun. No accident, then, that almost all our crop plants are sun tolerant. And no accident that wherever possible, we grow the shade tolerant ones in direct sunlight as long as they're sun tolerant as well. Neither shade-grown tobacco nor coffee provides a serious energy source for hungry humans, and black pepper doesn't either.

Anyone who has suddenly entered a forest from an area of full sunlight knows of the dramatic drop in light level. Fortunately, our visual systems adjust quickly, and while we know that beneath the canopy light intensity is less, we rarely retain that initial sense of just how much the canopy attenuates the light. That rapid adjustment holds real value for a running, hunting, and sometimes hunted animal, but it does mean that our judgments of light intensity can't be relied on. (The visual systems of many flying animals appear to use special tricks to accelerate adjustment to changes in light intensity. Presumably, navigation becomes a particular problem when flying from sun into shadow, and vice versa.)

The photographer knows it all too well—or did, before autoexposure cameras took care of much of the problem. Instructions for setting cameras without light meters usually suggested looking at the presence and intensity of shadows to judge the intensity of sunlight. One quickly appreciated that even "open shade" represented far less illumination than did direct sunlight.

Do it yourself: Anyone who has a camera that provides a readout of the diaphragm and shutter settings for a photograph—on recent models, it's on the LCD on the back—can figure out how much illumination changes from field to forest. At each location, just point the camera at a large piece of, say, brown wrapping paper. The photographer or automatic camera wants to admit about the same amount of light, done by some combination of adjusting exposure time and the width of the camera's diaphragm. Exposures, in fractions of a second, present no problem, but the diaphragm settings (f-stops) need some explaining. They form an odd scale, one chosen to indicate how much light will reach the light-sensitive surface (formerly the film) rather than simply the width of the lens opening. Halving the number in the f-stop, as from f/8 to f/4, means that four times as much light will strike a given area of the light-sensitive material. So going from one-sixtieth of a second at f/8 to one-thirtieth at f/4 increases the light input by a factor of 8, appropriate if the scene is eight times less well illuminated.

Note: "f" stands for the distance between the lens and the photosensitive surface. So "f" with a number under it means that so-called focal length divided by the width of the diaphragm in the lens. Putting in the "f" rather than some constant provides a proper correction for a shift from, say, a wide-angle to a telephoto lens: to get the same area-wise light on the sensor, the longer lens has to be wider.

The area of the two-portion opening on a spaghetti measurer is twice the area of the one-portion opening; the widths of the openings vary less, since area tracks not width but the square of the width. Same thing.

After writing out the instructions above, I followed them—at the moment I'm writing on an island north of Seattle, Washington, with lots of dense Douglas fir forest practically at my doorstep. Aimed at the ground in full sunlight, my camera reads f/6.0 at $\frac{1}{250}$th of a second; while aimed at fully shaded forest floor, it reads f/2.6 at a quarter of a second. Dividing 6.0 by 2.6 and squaring the result gives a 5.33-fold difference; $\frac{1}{4}$ divided by $\frac{1}{250}$ gives a 62.5-fold difference. 5.33 times 62.5 is 330. So that patch of well-shaded forest floor received 330 times less light than the nearby ground or the forest canopy.

Most of the light that no longer reaches the forest floor has been either absorbed or else reflected back upward by leaves. Since a deciduous forest blocks a lot less light in winter than in summer, we like to hike around our local forests in winter and early spring—both we and the wildflowers of the forest floor enjoy that huge improvement in illumination. Many understory plants do most of their growing and flowering before the trees overhead have leafed out.

The understories of mature temperate forests may be dark; tropical forest understories are a lot darker. The title of Joseph Conrad's great novel, *Heart of Darkness*, set in the Congo, alludes to just that. Sunlight, on average, is only a little brighter in the tropics—there's less than a twofold difference in intensity between a sun that's directly overhead and one that's inclined at 45°. (You can easily test that assertion with pencil, paper, and ruler.) What's different are the conditions favoring plant growth, conditions such as uniformity of temperature and continuous availability of water. When walking on the floor of a tropical forest, a person is dazzled when coming upon a place where a big tree has fallen and the sun penetrates the canopy directly. Photography becomes tricky—brights are too bright, darks too dark—as I found out when taking pictures at the Smithsonian lab on Barro Colorado Island, in Panama.

There's another difference between sunlight and what might be called shadelight. Direct sunlight amounts to a point source—turn away by more than a few degrees and the intensity drops sharply. Shadelight provides a distributed source, so intensity varies less with orientation. Leaves looking for illumination in the shade ought to be less fussy about their postures than leaves in the sun. (That is, unless direct sunlight so overwhelms the photosynthetic machinery of leaves in the sun that orientation just doesn't matter. But that doesn't seem to be the case, since the sunlit leaves of most broad-leaved trees do adopt a reasonably sunward posture.) Casual observation of leaves in the shade suggests that they're less fixated on orientation, but I don't know of a definitive study, and I cheerfully admit my predilection for self-deception when I have a hypothesis in mind. Objectivity doesn't come naturally, which is one reason we should distrust anecdotal evidence about anything.

How Light Drops Off

Light penetrates something that lets only a little get through in a somewhat counterintuitive way. One might imagine that two leaves, one beneath the other, would extract twice as much of the light falling on them as would one leaf. Not so, which brings up another generally useful bit of quantification. Imagine a leaf that removed just half the sunlight hitting it. Putting another such leaf below it means removing an additional half of half, or a quarter of the light.* The pair, in all, removes three-fourths of the light. Adding a third leaf would increase absorption to seven-eighths. Reality dwarfs these numbers, with a typical leaf transmitting only about 5 percent of the light striking it. So two leaves, one below the other, would transmit 0.05 times 0.05, or 0.0025—a quarter of 1 percent of what hit the top leaf. Now, leaves don't just turn off when not engaging in photosynthesis, even if that's the impression we get from some biology textbooks. Like animals, leaves respire; like animals, they do so all the time. As light diminishes, a point comes at which the oxygen production of photosynthesis is entirely reinvested by the leaves' own respiration—the compensation point. That happens at about 1 percent of full sunlight.[2] So the passage of light through two leaves will leave an intensity inadequate for the leaves' business. And that assumes full midday sunlight.

The way objects—both solids and solutions—absorb light makes a difference in how we go about our daily lives. You fill a cylindrical mug half full of black tea. The tea seems too strong, so instead of filling the rest of the mug with tea, you add boiling water. The fully filled mug, oddly, looks just as dark as the half-filled one did! No mystery when you think about it. Light had to go through just as much tea (never mind water) to get to the cup's bottom, bounce off, and return outward for you to see. Some particularly mundane photographs (fig. 2.6) show this teacup paradox.

Most trees have lots of leaves beneath the topmost ones. Why don't these trees get sensible about the situation and stop wasting resources making anything but a top layer or two of leaves? We know better than to put one solar collector in the shade of another. If the Beer-Lambert law is correct, then nature knows about it, just as she knows about all relevant

---

*The general statement is called the Beer-Lambert law of absorption by physicists and simply Lambert's law by chemists. It can be put as follows: Each layer of equal thickness absorbs an equal fraction of the light that passes through it. Note—"equal fraction" rather than "equal amount." That describes what's called an exponential decrease, and so the Beer-Lambert law can be written as

$$I = I_0 e^{-\alpha l}.$$

I and I_0 are the final and initial light intensities, respectively. e is 2.71828 . . . (the base of natural logarithms), α (lowercase alpha) is a situation-dependent absorption coefficient (how absorptive is a leaf), and l the path length (in effect proportional to the number of leaves).

FIGURE 2.6. The two cups at the top of this photograph hold tea poured directly from the same pot—the apparent difference in strength comes, as you'd guess, from the fourfold difference in depth. The two cups at the bottom may look the same, but they differ in strength. A quarter-cup of tea was poured into each; the right-hand one received three-quarters of a cup of water as well.

physical rules, whether or not they're in our textbooks. Evidently, she judges that despite the lower illumination, putting leaves beneath other leaves is still cost effective in terms of photosynthetic yield relative to the cost of construction, mechanical support, and ongoing maintenance. Explaining the paradoxical response of trees to the teacup paradox goes a long way toward aligning an idealized physical picture with the real world of trees and leaves.

Leaves Above and Below

Recall my measurement of a 330-fold reduction of light intensity between direct sunlight and that reaching the floor of a forest. I deliberately picked a dark place where almost nothing grew beneath the trees. In actual practice, that is, averaged over time and space, things rarely become so dire. When we account for leakage through holes in the canopy and the direct and diffuse light from them, a fiftyfold reduction might be typical.[3] This last figure says that trees intercept about 98 percent of the light striking a forest. Extracting that last 2 percent isn't economical for the trees that make up the canopy. It isn't worth the cost of making and maintaining the collecting equipment—at least relative to investment elsewhere in growth, production of acorns and pinecones, and so forth. Two percent is also satisfyingly close to the compensation point of around 1 percent, the point, again, at which photosynthesis no longer gives any net yield. Two percent also puts light intensity in the range in which competition from growing understory vegetation will be suppressed, so a tree needn't do better on that score either.

That light arrives at the forest floor as both diffuse and direct illumination. Diffuse light is skylight (as opposed to sunlight) that has come through holes in the canopy, as well as sunlight that has either diffused due to the atmosphere's particulates or hit leaves without being absorbed. Direct light that passes through the canopy goes by the name of sunflecks. At a given location, sunflecks have peculiar temporal characteristics. As the sun moves across the sky (sorry, Copernicus) different places in the canopy receive direct illumination, with the sizes of the holes in the canopy determining the duration of these shafts of light. In addition, wind puts canopy leaves in motion, generating very brief, pulsating sunflecks. Sunflecks move around at speeds vastly greater than any possible growth-associated leaf repositioning. They make investment in understory look a lot more reasonable.

There's another part of the story, another counterintuitive element that explains some everyday yet puzzling features of a forest. It's best posed as a question, one we've already met. Would a tree be better off making a single layer of leaves, each fully exposed to the sun and together blocking the sky, or would it be better off making a multilayered array of leaves?

Some trees are closer to the former leaf configuration, some to the latter, so the answer will most likely be situation dependent.

Some years ago, Henry Horn[4] gave an intriguing functional explanation of what we see in a forest. Two totally different phenomena provide the key. First, the sun has width—it's not perceived as a point source of light either by us or by leaves. And second, in full sunlight, leaves can use only about 20 percent of the light striking them—the photosynthetic system saturates.

Because the sun has width, the shadow of an object in the sun gets blurrier with distance, even in perfectly clear air. A fully shadowed area exists only closer than about a hundred widths of the object casting the shadow.* (Put another way, an object casts a complete shadow for only a hundred times its width—100 feet behind a pole 1 foot in diameter.) So a shaded leaf more than about a hundred widths below a sunlit leaf can't be fully shaded by that sunward one. If leaf width is 5 centimeters (2 inches), full shade ends about 5 meters (16 feet) below it. If the leaf is a pine needle half a millimeter across, then its full shade ends a mere 5 centimeters (about 2 inches) below.

The way photosynthesis saturates means that a shaded leaf far enough below the shading leaf (the umbrellar one, you might say) may receive light close to the saturation point. And that can happen without any fluttering of leaves in breezes or movement of the sun across the sky. So being below the top layer of leaves isn't as bad as it might seem. And—skipping to the bottom line—a multilayered tree can expose twice as much leaf area to light sufficiently intense to saturate the photosynthetic system of its leaves, compared with a tree having a single layer of leaves that completely shadow everything below.

But what about understory trees—or other plants? If the initial light level is low, then the monolayer arrangement of leaves is best. In effect, light no longer exceeds saturation by enough to warrant a multilevel strategy. The point at which monolayer and multilayer arrangements do equally well is a light level between 20 percent and 100 percent of full sun-

*The sun is 150 million kilometers (93 million miles) from the earth and 1.4 million kilometers (860,000 miles) in diameter. The ratio of the two numbers is 108. With a simple drawing, you can check that this ratio is the one relevant for the present argument about shading.

light—at 54 percent of full sunlight, according to Horn's model. Still, the monolayer arrangement, even if suboptimal for the individual tree of the forest canopy, at least discourages the growth of competitors beneath it. So, according to Horn, the development ("succession") from an abandoned field or blowdown to a mature forest ought to have multilayer trees early on, with their capacity for rapid growth, and then monolayer trees when mature, when competition can be suppressed but no longer outgrown. Does this correspond to what we see in nature? He believes so, and if I'm not in a particularly skeptical mood, that's what I see when walking through forests of varying successional stages in central North Carolina.

More consistently persuasive is something I see in the stand of fairly old pines in my front yard. They're multilayered in just the way trees with very narrow leaves ought to be. No use spreading the underlayers over an extensive vertical dimension. That would take additional structural material, and the hundred-diameter point will be reached, as we saw, in centimeters or inches. Instead, they have clusters of needles splayed out in arrays near the top, quite different from the leaf arrays of almost all our broad-leaved trees, which extend quite far down from the top. Of course, Horn's analysis rests on a severely idealized model. That brings to mind words from John Burns, to whom the designation poet "laureate" ("crowned with laurel leaves") is especially apt:[5]

> A model in its elegance
> Is better than reality
> Its graphical simplicity
> Denotes a rare intelligence.
>
> The simple graph incites the wrath
> Of field men who, half undressed,
> Go rushing out to start a test
> Which culminates in aftermath.

Might the arrangements of leaves on trees provide models for how we should array our solar collectors? The vertical spacing needed for collectors of the size we find economical to build would be impractically great. Making many tiny collectors instead of a few large ones would have to incur an almost trivial incremental cost. Besides, leaves have other things to worry about than do the designers of solar panels. Full sunlight may be not just wastefully intense but, as I'll argue when we return to the problems of leaf heating, too much of a good thing. Between that and all the other incidentals of tree-ness—compensation point, the way tree height fails to approach the sun, the routing of their systems of vessels, the branching systems that allow for growth, and so forth—the way trees place their leaves gives precious little guidance to our own solar collectors. Here and elsewhere, unless we understand the full multifactorial story, nature will more likely mislead than lead.

In any case, leaves on a tree do not neatly divide into "always in direct sunlight" and "always shaded by others." Even minimal wind, changing solar position, and so forth enable lower leaves to be intermittently exposed to direct sunlight. Still, the extremes retain reality. The topmost leaves remain in full sunlight, while for many understory plants, no leaves see full sunlight at any point during the day. In the piedmont of North Carolina, where I live, we have both broad-leafed hardwood forests and needled softwood forests, the latter mostly made up of loblolly pines. The pines reseed themselves naturally (and are also plantation-grown for timber), but they do so only in clearings provided by abandoned fields, roadsides, and so forth. The pines cannot establish themselves in the shade of other trees, including, oddly, their own adults—they're incompetent as understory inhabitants. So an abandoned field first becomes a pine forest but later—the better part of a century later—develops into a hardwood stand with no pines left at all.

Why Put Leaves High off the Ground?

One last item, something mentioned as a convergence in the previous chapter, something so familiar we forget to ask why it might be so. Trees grow tall. The earth has supported trees and treelike plants almost since the appearance of plants with internal piping—vascular tissue—back in the Devonian period 400 million to 350 million years ago. The Carboniferous period, 350 million to 300 million years ago, saw extensive forests of huge trees, some well over 35 meters (115 feet) tall, including the prolific plants from which most of our coal deposits derive.[6] We might guess that tree height should increase over evolutionary time, but that turns out not to be the case. That a few trees at present exceed 100 meters (330 feet) needs to be viewed in context. Forests much over 35 to 40 meters (115 to

130 feet) are uncommon at present, and we have only a remote chance of discovering measurable fossils from exceptionally tall specimen from any past era—"exceptionally" means just that.

Nor are all present and prehistoric trees parts of the same lineage—far from it. As chapter 1 noted, treelike forms have independently evolved from small plants in a large number of instances. These treelike plants not only look different from one another, but their internal structures and growth habits differ widely as well. Even at present, within our own slice of planetary history, we have forests of both softwoods and hardwoods, to say nothing of palms and bamboos, these latter remarkably different from the first two. The more familiar softwoods and hardwoods grow in diameter as well as height; the palms and bamboos grow only in height, each plant becoming relatively skinnier as it gets taller.

In every lineage that has independently come up with treelike plants, a variety of species achieve great height. That appears to me to be the height of stupidity. Again, extra height can't provide greater solar intensity by shortening the distance to the sun—what's even a hundred meters when the sun is over 90 million miles away? Natural selection can't possibly capitalize on such trivial gain. We're looking at, almost surely, an object lesson in the limitations of evolutionary design. The early bird gets the worm; the topmost leaf gets disproportionate (the argument above) access to light. Although an individual tree might disport its leaves in some rational array, a forest is stuck with irrational competition among its trees. Yes, under some circumstances competition can be ameliorated by common advantage or tactical cooperation, but trees apparently don't meet the requirements for group intelligence, at least with respect to height.

A trunk-limitation treaty would permit all individuals to produce more seeds and to start producing seeds at earlier ages. But evolution, stupid process that it is, hasn't figured that out—foresight isn't exactly its strong suit. If you wish, you might regard the wonderful height of trees, our source of wood for fuel and construction, as evidence for a crude and wasteful process as opposed to properly rational design.

Again, the tallest trees have remained largely unchanged in height since treelike organisms first became common. That tree height hasn't changed implies at least four things. First, it suggests that tree height isn't unlimited. Maybe some factor or combination of factors sets a particular upper bound (something I'll defer more talk about for now). Second, it implies that the supply of whatever resources are crucial in plants that go for height hasn't changed much over the past few hundred million years. Greater photosynthetic efficiency, for instance, might permit the point of diminishing returns to occur at a greater height. So one surmises that the basic biochemistry of the process was well set before trees made the scene—and the biochemical similarity among the great diversity of present plants provides strong confirmatory evidence. Third, we can similarly surmise that wood has been wood, as far as its relevant mechanical properties go, for this whole time as well. Again, similarities of genes and biochemistry give the same picture, if a bit less persuasively. Although cellulose may remain cellulose, woods are quite diverse at higher structural levels. No great surprise—after all, a fluff of cotton fibers and the twisted cotton thread we make from them have exactly the same composition. Finally, changes in atmospheric composition, especially of the key resource, carbon dioxide (CO_2), which we know has varied, seem to matter little for height—whatever CO_2's role has been in setting the rate at which plants produce vegetative matter.

The height of trees will come up repeatedly in chapters to come. For now we'll leave further ramifications, as it were, up in the air.

NO LEAF CAN DO WITHOUT three things: light, carbon dioxide, and water. Light we worried about in the previous chapter; this one and the next look at the supply of carbon dioxide as well as gains and losses of water vapor.

Although a leaf lives in a world that doesn't lack for light, carbon dioxide, CO_2, is scarce stuff. In our terrestrial atmosphere it currently makes up just under four-hundredths of a percent—0.04 percent, or 1 part in 2,500—of the overall mixture. For comparison, nitrogen, which we rarely think about, makes up about 78 percent; and oxygen, critical for nearly all of us animals, constitutes about 21 percent. Even argon, no household word, accounts for nearly 1 percent of the atmosphere, twenty-eight times more than CO_2.

Perhaps leaves need a lot more oxygen-sucking, CO_2-belching animals—respiring animals have long been their most reliable suppliers of this gas. Of course, most animals are herbivores, as demanded by a basic ecological rule for energy flow, so we're not exactly an unmixed blessing for plants. Or maybe what leaves need is just what's happening right now: a gradual increase in the CO_2 in the atmosphere, almost entirely the result of massive burning by our species of the carbon contained in fossilized vegetation. After all, we're helping those dead plants recycle their remains. Lots of recent experimental work shows that increasing the CO_2 concentration around a plant usually stimulates growth. But when delivered as a change in atmospheric composition, the increase also causes global warming. So we might advise a plant not to wish for CO_2 enrichment. It might get its wish and find that the resulting climate shift lets some other kind of plant take root and put it out of business.

Ironically, the scarcity of atmospheric CO_2 that's facing the leaf marks the legacy of forebears who almost completely depleted the atmosphere's supply of the precious gas, the carbon source for 99.99 percent of life. From the leaf's point of view, altogether too much of the earth's carbon is locked up in water as carbonic acid, and in stone as calcium carbonate and similar compounds—in the fossil forms that we mine and burn, in the clathrates of the tundra and seabed, and, of course, in other vegetation, alive or of recent demise. By comparison, all the animal life on earth sequesters a trivial amount at any moment. Even though animals process the stuff at a pretty decent rate, there just aren't all that many animals on earth.

Getting water, the other input for photosynthesis, poses a lesser problem. Air contains far more of the stuff, even when the humidity is low. For instance, when the air temperature is 25°C (77°F) and the relative humidity is just 25 percent, about one-half of 1 percent of the air is water vapor, a concentration more than sixteen times as great as its charge of CO_2. Moreover, diffusion moves the water vapor, H_2O, a lot more rapidly, so water's relative availability is even greater than that factor of 16 implies.

Moving Gas Diffusively

A leaf can't get CO_2 by inhaling enriched air and exhaling depleted air—rapidly moving parts such as muscles and the like don't exist in its armamentarium of devices. It can't reach out beyond its cells and snag or entice molecules of CO_2. Since biology comes up wanting, the leaf has to rely entirely on physics. The best it can do as it sits there on its stem is to arrange itself to facilitate whatever physical processes present themselves. Two quite distinct processes can help, and it takes advantage of both, most often simultaneously. They're the same ones that, with our very different equipment, we take advantage of when we breathe. Most of us misunderstand not only how both processes work but the sharp distinction between them. And both loom large (and largely ignored) in our everyday lives. One is diffusion; the other goes by such names as bulk flow, wind, and convection.

First, diffusion, a stranger business than you ever imagined—or than you were taught in science classes. I'll spend few pages on diffusion, in part because of its counterintuitive oddity, in part because misconceptions abound, and in part because of its transcendent importance to life. At this writing, the article on molecular diffusion in *Wikipedia*, usually a website fairly good on technical matters, consists of a series of howlers, with references at the end to yet more mythology. (But the article on diffusion coefficients looks fine.) I once published a diatribe aimed at just this mythology with—fifteen years later—little effect, even within the target audience.[1]

At some time or other you probably heard that the molecules in gases and liquids, and even in some solids, move around randomly, entirely without goading by any outside agent. You may even have heard that the rate of such wandering depends on the temperature. Or it may have been put the other way around: temperature amounts to a measure of the rate of that spontaneous, orderless motion. In any case, I mentioned in passing the association of temperature and molecular motion back in chapter 1.

At any temperature above absolute zero, each mobile molecule moves in a straight path until it hits another molecule, whereupon it collides, bounces off, and heads in another direction. That's the basis of diffusion. You might think that random movement with frequent change of direction would get you, or a molecule, nowhere. That's both true and untrue, depending on how you look at things. On the average, you, or a molecule, get nowhere—after all, with movement in every direction, things average out. But—and here's the crux—as time passes it gets ever less likely that at any particular time you'll be back at the point in space where you started.

Think about this peculiarity of wandering. Take one step from the start in a randomly chosen direction; the next step might well bring you back to where you started. Take ten such steps, and the result of the next one—or even ten—is far less likely to return you to that point. Randomness isn't something we think about very well. If a person is asked which of several patterns of dots on a page is most likely to be random, the person will almost inevitably choose one that's much too uniform. But a computer does the job well, so I've asked such a properly unbiased (since mindless) machine to produce tracks of random wanderers; figure 3.1 is one result. We could, I suppose, approximate the process with a group of people, each carrying a roulette wheel marked with possible headings of their next steps.

On the average, then, you go in no direction that you can anticipate. But on the average you will move away from your starting point. Now, with molecules, that "on the average" takes on great importance, enough so that the specific behavior of any one molecule at any time interests

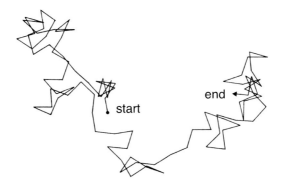

FIGURE 3.1. A computer simulation of a two-dimensional random walk. Here at each of a predetermined number (100) of steps, the walker picks a direction at random and then moves a standard distance in that direction. Any computer program (Basic, C++, etc.) that includes a random number generator and a screen driver can generate such patterns.

Do it yourself: You might try a rough version of such a random wander. An easy version gives you a choice of only four possible directions and a standard step length. With four possibilities, a pair of coin-flips will provide a properly random choice. Thus head-head says go north, head-tail says go east, tail-tail means south, and tail-head means west. Not dropping the coin provides the only challenge in this admittedly unexciting game. It might be fun to try it with elementary school students—who ought to be introduced to random processes anyway.

us very little. The reason is simple: we're so much larger than individual molecules that we care almost entirely about their collective rather than their individual behavior. Even cells greatly exceed molecules in size, a point that's insufficiently appreciated. When I taught introductory biology, I'd emphasize to the class that there are about as many molecules in an (animal) cell as there are cells in, say, a cat. Molecules, even macromolecules, are damned small. To our eyes or to any ordinary measuring devices, a random distribution of dye molecules in water looks completely uniform. So we call the concentration uniform after we stir the sugar in our coffee, even if on a molecular level it's anything but uniform.

Thus diffusion means that stuff moves, if in this peculiar way. If what we're looking at happens to be present in uniform concentration as viewed on our scale, like dissolved sugar in a cup of coffee or a roomful of the constituents of air, the movement has no consequence in terms of transport. But if something isn't uniformly distributed at the start, it gradually approaches uniformity. Strange but true—an undirected process now appears to have direction. Stranger—the process seems to know where the diffusing material is in short supply. We can restore confidence in our sense of reality with the following analogy. Imagine two rooms, equal in every way, with an open window between them but that are otherwise closed. In one a whole bunch of flies buzz around; the other room is initially empty. Of course, flies will fly from the fly-filled room into the empty room, even if they have no knowledge that the room is empty. As the initially empty room accumulates flies, some will fly back. But the net rate of back flying will not reach the rate of flying in the initial direction until the concentration of flies in the two rooms approaches equality.

With a permutation of our model, we can make a room accumulate flies perpetually. Imagine two changes in the previous model: replace one of the rooms with the great outdoors, and line the second with old-fashioned flypaper. An unlimited supply of flies can move into second room; few of the flies will move out before being trapped by the flypaper. At least that's what happens if we do the kind of simplifying tricks that ideal models legitimize, like making permanent, unclogging flypaper and worlds of unlimited size hosting uniformly distributed flies. Simplified, perhaps, but the model resembles things both real and consequential enough, so it provides a useful conceptual aid.

To put the flight (or plight) of these flies and the wandering of molecules in general terms—we do so love our generalizations—net or average random movement is directed (paradoxically perhaps) from regions of higher to regions of lower concentration. Put in a leaf's context, if its cells use (= trap) CO_2, then CO_2 will move from the atmosphere to those cells by simple diffusion. It will do so entirely without the cells of the leaf bestirring themselves (or stirring the air) in any way at all. And the CO_2 will continue to diffuse inward as long as the cells (= flypaper) continue to use it. Of course some molecules of CO_2 will go the other way, drifting away from rather than closer to the cells. But we can sweep that matter under the rug by (again, both cells and ourselves being large) looking only at overall, average movement. Via diffusion in gases and liquids (collectively "fluids"), nature provides a wonderful free ride.

(Thermal energy, expressed as random molecular movement, ultimately drives diffusion, in case you're imagining a truly free lunch. In the end, diffusion is a direct manifestation of the second law of thermodynamics, talked about in chapter 1. Things become more random: the location of a fly somewhere within two rooms is less certain than its location within only one room. And the temperature must be above 0 K, or molecules won't feel compelled to move—in fact, as you now understand, saying that about temperature amounts to a circuitous assertion.)

The central relationship that describes diffusion is Fick's law. It relies for a model on a slab of material through which some kind of molecule diffuses. In effect, it replaces the window between the rooms in our analogy with a short tunnel. That provides a travel route for diffusion, retaining the limitation of overall movement to a single direction (or a pair of opposite directions, if you prefer to think of individual molecules). Incidentally, Adolf Eugen Fick (1829–1901) was a physiologist rather than a physicist—diffusion long ago loomed large in biology.

As Fick's law puts it, the amount diffusing per unit time equals three things multiplied together.* The law starts with the difference in concentration between the sides of the slab. A greater concentration leads directly and simply to a greater net (repeat, net) amount diffusing per unit time—whether flies or molecules per unit volume. Fick's law then divides that concentration difference by the length of the path—thickness of the slab or length of the tunnel—because the longer the path, the more time the process will take. That computation gives the first factor,

---

*Assuming (to avoid calculus) a steady, unidirectional process as well as a minimal concentration gradient that's uniform along a very short length of the path, Fick's law (sometimes "Fick's first law") can expressed as

$$\frac{m}{t} = -DS\frac{C_1 - C_2}{x},$$

where m/t is the amount (mass) of material moved per unit time, t; D is the diffusion coefficient, S the cross-sectional area of the path, C_1 and C_2 are the concentrations (mass per volume) of material at the source side and the sink side, respectively, and x is the path length. The minus sign reminds us that net diffusion goes from higher to lower concentration. From this formula, swell equations can be derived for every manner of geometric and temporal situation.

called the concentration gradient. (Gradients are as important in physics as they are to any runner.) Second comes the cross section—the face area of the slab of material or the bore of the tunnel. A wider path means more material passes per unit time. The third and final multiplier accounts for the kind of molecule that's diffusing together with what it's diffusing through; it's called the diffusion coefficient. In practice we look up values of the coefficient from ones tabulated in handbooks, either on paper or online. Smaller molecules diffuse more readily than larger ones, all molecules diffuse very much more readily through gases such as air than they do through liquids such as water, and diffusion through solids goes even more slowly.

The differences among gases, liquids, and solids matter greatly both for leaves and for every other biological system that involves at least two of these three phases of matter.* Diffusion coefficients run around ten thousand times greater for movement in air than in water; coefficients in living tissue run around half those in water.

Speed and Distance

Mention of the diffusion coefficient brings us to the central issue in any quantification of diffusion: how fast and how far. The situation is as odd as the business about getting directionality out of a random process, so we'll ease in gently and intuitively. The diffusion coefficient is a kind of rate, with higher values when diffusion is speedier. But proper rates are written as distances per unit time, such as miles per hour or meters per second. By contrast, diffusion coefficients have dimensions of distance squared per unit time.† Rates, yes, but not in exactly the familiar sense— again, we're not talking about a discrete entity moving in some consistent or predetermined direction.

---

*The three phases of matter can perhaps be best envisioned as follows. Solids have both size and shape; liquids have size but not shape; gases have neither size nor shape. For present purposes, most of the differences between liquids and gases are just quantitative ones—numbers rather than natures—which is why we so often use the generic term *fluids* for either or both. Proper solids are something else altogether.

†Dimensions are variables such as length (or distance), time, energy, and so forth. The scales we put on them are units, such as (staying with these three) meters, seconds, joules, et cetera. Or feet, hours, calories, and the like.

The difference in dimensions hints at what's happening. For normal, steady travel, going twice as far takes twice as long. Or doubling the rate of travel means a given trip will take half as long. A truck going back and forth will transport only half as much if it has to go twice as far on each trip. For net diffusive transport, though, a doubling of distance yields not a twofold but a fourfold increase in travel time, or a quartering of the rate of transport across the distance. Even if the diffusion coefficient isn't a rate in the familiar sense, it's at least a kind of rate—higher values do mean speedier transport, more transport in any given time.*

Why quibble about this quantitative queerness? Fick's law implies that diffusion deteriorates dramatically as a mode of material transport as distance increases. It may be free, but it's a ticket to nowhere very distant unless the system is very, very patient. Say some given amount of a chemical can travel a hundredth of a millimeter (about the diameter of an animal cell) in a second. To go a full millimeter, the same amount of transport would take not a hundred but no less than ten thousand seconds, almost three hours. By that time the predator would have long since pounced. Going a meter (lung-to-thumb distance, for instance, relevant for oxygen) would take over three centuries. Applying the same scenario, diffusion from root to leaf within a tree would take longer than the time between the end of the Neanderthals and the present. Yes, plants do live their lives at slower paces than we do—but obviously not quite that slowly. Again, the big point: diffusion may be both cheap and effective for very short distances or very long times, but it's a fool's errand otherwise.

Now the diatribe. You may have been subjected to a certain classroom demonstration of diffusion. An instructor pours out some perfume at one end of the room, and pretty soon everyone gets a whiff. (Why anyone should bother doing this I do not know; we're all familiar with the perverse, perfidious, and pervasive power of flatulence.) The demonstration,

*Examining the dimensions of Fick's law tells us that diffusion coefficients must be expressed as length squared per unit time. Volume has dimensions of length cubed, area dimensions of length squared, and concentration has dimensions of mass per unit volume, Mass and time, on the left-hand side of the equation, need no reduction. Do the simple algebra—D comes out as $L^2 t^{-1}$. The exercise illustrates a down-to-earth use of what might look like an arcane abstraction. Not so arcane, really: equations that properly describe aspects of the real world must have the same dimensions on both sides as well as the same numerical values.

be assured, is a 99.99-and-then-some percent fraud! Diffusion becomes consequential only within a fraction of a millimeter of the perfume's surface and within a similar fraction of a millimeter of the living tissue lining your nasal passages. The rest is wind—*flatus* in Latin—both as a physical agency and in English as an explanation. If the air movement in the room could in some way be stopped, diffusion would indeed bring the odorant to you—in a mere month, as Howard Berg calculates.[2] I do love demonstrations as teaching tools, but contriving a decent demonstration of diffusion in air defies my creativity. I note with some self-justification that the famous separation of uranium isotopes by gaseous diffusion during World War II depended on an exotic porous barrier through which the uranium hexafluoride gas passed—the stuff wasn't diffusing in an open container.[3]

Demonstrating diffusion isn't easy, even in water. Common bits of show-and-tell such as the spread of dye from a crystal depend on water currents, that is, on flow, far more than they do on true molecular diffusion. If you can see any order at all, such as swirls or waves, you're looking for sure at such bulk fluid flow, not at diffusion. Keeping liquid water sufficiently still to rule out flow turns out to be no mean task. Worse than just misattributing what's happening, such demonstrations give us an entirely wrong impression of diffusion's efficacy as a mode of mixing things over substantial distances within short periods of time.

Not easy doesn't mean that a demonstration isn't possible. The key consists in keeping the liquid from the slow self-stirring induced by even minor temperature variation or persisting from the initial filling of the container. One way to observe diffusion over a visible distance in a reasonable time involves adding a small amount of a gelling substance to stabilize the liquid—in effect making it into a semisolid, which is what I did to generate figure 3.2. Another proper demonstration of diffusion depends on a liquid system with a very high viscosity. Viscosity—more in the next chapter—represents a fluid's (yes, even gases have viscosity) unwillingness to flow, so lots of it is just what we want. But it takes a bit more sleight of hand than does using gelatin or agar.

Over any short time span, demonstrations that show real diffusion are persuasively unimpressive. Bottom line: diffusion over distances on our personal scale proceeds so slowly that even modest bulk movement completely eclipses it. Conversely, on a microscopic scale no transport

FIGURE 3.2. Penetration of a colorant into gelatin at room temperature. Knox gelatin was made up at double the concentration suggested on the package; the colorant is ordinary food coloring (McCormick's Red 40 and 3) diluted severalfold with water.

Do it yourself: Half fill a cylindrical, transparent glass or beaker with colorless gelatin, made up perhaps twice as concentrated as the package directs. Let the stuff gel in the refrigerator, with the container covered with plastic wrap. At the start, very gently pour a fairly strong solution of food coloring down the inside of the container and over the gelatin. The interface will be sharp. Then observe occasionally over the next few hours and days—the colorant should gradually penetrate the gelatin and fuzz the boundary between the two substances. You might even try mixing several colors. More rapid diffusion of colorant consisting of smaller molecules will lead to a certain amount of separating. Chemists have long made use of the phenomenon to uncombine mixtures.

Do it yourself: Sugar syrups have very high viscosities, up to around ten thousand times that of pure water, and they come from the nearest grocery in both clear and colored form. To a glass jar filled with clear syrup (Karo or other brand), gently add just a little dark molasses with as little disruption as possible. The two substances won't layer all that neatly, but they won't immediately mix either, and the motion from

process surpasses it in importance. A great deal of biology happens on just such a scale. In the animal world, diffusion dominates as a transport mode within cells, while over larger distances forced flow does almost the entire job. (That may explain why most animal cells are the size they are.) In the world of our leaf, the distinction turns out to be less crisp. That's not because the leisurely life of plants permits diffusion to be effective over longer distances. It's just that typical plant cells, much larger than animal cells, have their own individual internal pumping systems to force fluid around. So the rule is general: for movement over anything beyond a few micrometers, nature has had to invent all manner of pipes and pumps. And she has had to pay the attendant energy cost. That cost isn't trivial. You expend about 10 percent of your resting metabolism just pushing blood around—to bring the issue right to where you live.

Going Inside

To carry the story further, we have to look at what happens to CO_2 inside the leaf. Back into the picture comes structure, of little concern since talking about the arrangements of leaves on trees. Figure 3.3 gives a view of the inside of a typical leaf, in this case a thin crosswise slice.

It puts a few (a very few) of the botanist's names on things, names we'll need as we go.

Looking at the cross section with our peculiarly physical prejudices, several things jump out right away. Leaves have a lot of air inside. Within that lower layer of irregular cellular columns, the "spongy layer," they have a continuous (or nearly continuous—veins interrupt it) set of air passages. But their upper and lower skins (epidermis) limit connection between those internal passages and the outside air. One (most often the lower) or both surfaces are perforated with holes, or stomata (meaning "mouths"; the singular form is stomate or stoma)—small, lenticu-

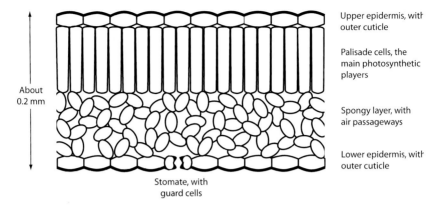

About
0.2 mm

Upper epidermis, with
outer cuticle

Palisade cells, the
main photosynthetic
players

Spongy layer, with
air passageways

Lower epidermis, with
outer cuticle

Stomate, with
guard cells

FIGURE 3.3. A severely idealized cross section of the blade of a leaf. The stomate shown should be thought of as leading into an interconnected set of air passageways amid the feltwork of cells in the leaf's spongy layer.

lar (lens-shaped) apertures. In the aggregate, stomatal openings represent about 1 percent of a leaf's overall surface. The width of each of these stomatal openings (maximally about a fiftieth of a millimeter, or 20 micrometers) can be adjusted by the inflation and deflation (with liquid) of a pair of cells, one on each of the stomate's sides. Recall the difference in ease of diffusion between gaseous and liquid systems, the ten-thousand-fold difference in diffusion coefficients. Clearly that internal air greatly aids the passage of gases between the atmosphere and the photosynthetic cells. (The air-filled tracheal pipes of insects take similar advantage of the greater speed of diffusion in gases.) And the stomata, those limited (and controllable) connections with the outside world, just have to be a crucial functional feature.

Considering the scarcity of CO_2 in our atmosphere, both the smallness of the apertures and of the fraction of the leaf's surface they represent seem distinctly odd. Diffusion demands a concentration difference and depends on the cross-sectional area of its path. The tiny cross-sectional area of stomata should limit penetration of CO_2, which doesn't sound right. For supplying the inside, lower entry area demands a greater concentration difference between outside and inside, and there's already precious little CO_2 outside. So the stomata ought to provide a worse bottleneck than the neck of any ordinary bottle. Oddly enough, fully open stomata don't particularly reduce inward diffusion of CO_2 below what a completely open surface would allow. Here Fick's law has to be put aside—or at least any calculation must take account of the peculiar geometry of the leaf.

A stomate forms a channel that's too short relative to its width for the law to apply in its simplest form. How the concentration changes immediately beyond the stomate, inside and outside, becomes more important than how it changes from one end of a pore to the other. Now, leaves have lots of stomata, and most have their stomata fairly close together. An open area much above that 1 percent of overall surface area buys little additional diffusion—the external concentration gradients just merge.[4] At the same time, closed stomata do shut down exchange, as one would expect. So having lots of tiny stomata that cover only a little of a leaf's surface provides control of diffusion from zero to a level close to what would happen if the whole surface were open for diffusion. That makes these tiny, adjustable openings remarkably effective for controlling the interchange between the external atmosphere and the air spaces within the leaf. Truly an elegant design.

Gas Exchange: Leaves and Ourselves

Like leaves, we're critically dependent on an atmospheric gas, in our case oxygen rather than carbon dioxide. If humans or most other mammals move from near sea level to some high altitude, breathing becomes troublesome. No mystery: at high altitudes the air contains less oxygen, not as a fraction (still 21 percent) but because of the lower atmospheric density and pressure. Lower density and pressure mean a lower absolute concentration of every atmospheric gas. Incidentally, the lower overall pressure itself presents no problem. We notice it only when our effervescent drinks get fizzier—noticeable, unless I'm self-deceptive, when a can is opened by a flight attendant at altitude. (Commercial aircraft may be pressurized, but not to full sea-level atmospheric pressure.)

Similarly, we expect plants to grow less well at high altitudes, given

Do it yourself: The next time you're in a plane that's about to take off, inflate a small spherical balloon and measure its circumference, perhaps using a piece of thread and a tape measure. Make the same measurement after the plane has reached cruising altitude; the change in volume should be just the inverse of the change in pressure. (Volume, bear in mind, varies with radius or circumference cubed, so the change in circumference won't be all that great.) You may have noticed that inflatable neck pillows get firmer when the plane ascends— same thing, if harder to measure.

the lower concentration of CO_2 in the local atmosphere and the general scarcity of the stuff at any altitude. We're all familiar with tree lines and the sparse vegetation near mountaintops. Right? No, as it happens—the analogy with our breathing misleads.

Why do we experience trouble at altitude? For one thing, we have to pump a larger volume of air through our respiratory passages. More important, even with more pumping from our hearts, we still have lower absolute levels of oxygen in our lungs, the oxygen that's available to diffuse into our blood and get picked up by our hemoglobin. So we can't fully load our hemoglobin with oxygen, and we have to pump more blood to get the same amount transported. It's no surprise that the system hits capacity at a lower level of exertion.

What about a leaf's access to CO_2? As with oxygen, at high altitudes its concentration in air remains about the same relative to other gases, while its absolute concentration goes down with atmospheric pressure. Access, though, depends on diffusion rather than on pumped bulk flow. Furthermore, that diffusion happens in a gas, air, rather than in a liquid, blood. And a funny thing happens when gases diffuse at lower ambient density and pressure.

Think again about the physical basis of diffusion, that random walking of molecules. If density drops, then those molecules go further between collisions, and therefore diffusion goes better. Put formally, the diffusion coefficient increases. So just as the concentration difference between the outside air and the cells within a leaf decreases, the diffusion coefficient of CO_2 increases. Diffusion coefficients change almost exactly the opposite way that density changes, so the balance is nearly perfect. Plants may have trouble at high altitudes, with colder temperatures, less reliable water sources, high winds, steep slopes, poor soil, and the like—but reduced access to CO_2 need not worry them.[5] That's unlike our situation, but not unlike most insects most of the time. As mentioned earlier, insects transport oxygen in gaseous form right out to their tissues rather than carrying it in blood. Yes, some, mainly adults during flight, pump air through their tracheal systems. Otherwise, as in leaves, diffusion does the bulk of the job, and altitude should make little difference to the effectiveness of the system.

OUR WORLD WOULD BE far less pleasant if the air around us wasn't in continuous motion—even what we think of as still air still moves, just too slowly for direct perception. Diffusion plays a role when any organism exchanges material with the outside world, when any cell exchanges material with its surroundings, and mostly when material moves around with cells. But, as the previous chapter was all about, it works well only in small-scale systems; it fails badly for any systems in which anything has to be moved for any distance perceptible to any of us. So how do living things manage, the ones that are neither very small, nor very flat, nor threadlike? They inevitably augment diffusion with bulk fluid movement—the ubiquitous ambient wind and water currents as well as pumped flows of air, blood, and lymphatic and other fluids. Unsurprisingly, the exchange of gases between the cells of leaves and the atmosphere depends on bulk flow as well as diffusion. That gives us a whole new bag of physical phenomena to explore: the world of fluid dynamics.[1] As we'll see, the operative rules for that business apply to other aspects of a leaf's life besides gas exchange—and to lots of things in our own everyday lives.

Moving Gas by Flow

At least flow moves material in the familiar manner. Once again, distance covered goes up in proper proportion to time spent traveling. Not that the flows of fluids don't have their share of counterintuitive subtleties and urban myths. We might ease in with a couple of surprisingly trustworthy simplifications. First, flows of air—winds—and flows of water—currents—behave pretty much the same way. Unless we have to deal with air-water interfaces and surface waves or with some kinds of heat transfer, the rules for one work about as well for the other. Yes, water's eight-hundred-fold greater density makes for greater forces, but that's just a quantitative, not a qualitative difference, no more than one simple factor in most formulas. And second, the density even of air, much less water, doesn't depend on how fast it flows, at least until one approaches the speed of sound. Altitude, temperature, depth underwater—all affect air density, but how fast air moves ordinarily does not.

Other things complicate the picture. Flows of both liquids and gases come in two basic types, *laminar* and *turbulent*, a pair of intuitively informative words. One way to describe the distinction may sound devious at first, but with a little further thought ought to sit well enough. In laminar flow, what's done by the fluid as a whole (or a fairly large chunk of it) amounts to a large-scale version of what each little bit of the fluid does. The whole, in short, does little more than any part. By contrast, in turbulent flow what's done by a large chunk of the fluid represents only a kind of statistical summary of the behavior of its parts. In a close-up snapshot, a turbulent flow looks like a messy mass of swirling eddies. Put another way, if a fluid is flowing laminarly north, then every little bit is doing so. If a fluid is flowing turbulently north, little bits may be going every which way. Marching versus milling, we might say.

Water may rush turbulently downstream in some channel, but introduce a bit of colorant somewhere, and the dye rapidly spreads across the stream—showing that there's lots of cross-stream movement within the overall body of water. This means that turbulent flow inevitably mixes a fluid as it transports a fluid. Laminar flow needn't mix the fluid, and if it does, it doesn't mix it nearly as much. Flows augment diffusion, but turbulent flows do so much more powerfully. Figure 4.1 makes that contrast.

Fluids—again, both gases and liquids—share a special physical property, the one that defines them as fluids, called viscosity. Viscosity might be described as an unwillingness to move unless pushed or pulled by some energy-supplying agency. But that's not quite good enough. That description fits mass equally well. Masses resist ordinary pushes or pulls, while viscous fluids in addition (they do have mass) resist pushes and pulls that change their shapes.* The coffee only seems to drop effortlessly into a cup

---

*Viscosity ("coefficient of viscosity," sometimes) is a measure of a fluid's response to shear. Neither solids nor fluids like to shear, but they react—or require force to shear—in fundamentally different ways. Shear amounts to deformation; in two dimensions, a rectangle becomes a parallelogram. The solid, remembering its original shape, wants to snap back; elasticity puts numbers on that. The fluid, with no such memory, just cares how fast it's sheared; viscosity is the relevant measure of that rate of shear. Put more formally, shear stress (τ, lowercase Greek gamma), a force per area,

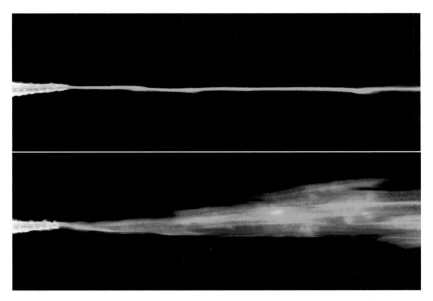

FIGURE 4.1. Laminar (*above*) and turbulent (*below*) flows. Here I've slowly injected fluorescein from the end of a piece of drawn-out catheter tubing into a stream of moving water in the flow tank (or flume) in my lab. The stream moved slowly in the top picture and more rapidly in the bottom one. The eddies of turbulent flow make fluid move sideways as well as downstream, blurring the dye track.

of any shape, so you don't notice viscosity in action. More viscous syrups take a noticeable amount of time to do so—and they take more force to stir them, in effect to change their shapes. Put a little more strictly, a fluid's viscosity is a measure of what it takes to slide one bit of fluid across another or else a bit of fluid across a surface. Newton's first law (which you may recall; don't worry if you don't) still applies, so if all bits are moving in the same way, it takes no force to keep them doing so. Where viscosity has a major effect, a flow will be laminar—it keeps the bits of fluid

equals the product of the coefficient of viscosity (μ, lowercase Greek mu) times the rate of shear:

$$\tau = \mu \frac{\Delta v}{\Delta y}.$$

Rate of shear is way the speed of flow changes (Δv) with distance across the flow (Δy), together the cross-flow speed gradient. We're following the convention that the x-direction is that of the flow, while the y-direction is the cross-flow direction, at a right angle to it.

marching in lockstep. Where viscosity plays a lesser role, a flow will break out in a rash of turbulent eddies as the bits mill about. On our perceptual scale in the outside world, almost all flows are turbulent, while within ourselves, meaning mainly blood flow, almost all motion is laminar.

Now, here comes an odd business. The size of an object or creature commonly determines whether it experiences a truly turbulent flow or just a laminar flow that varies in speed and direction from moment to moment. What matters is its size relative to the size of the eddies in the turbulence. Because it's smaller than the smallest eddy, a tiny planktonic creature in a turbulent stream or ocean experiences turbulence only indirectly. Conversely, turbulence defines the world of a tree in a storm; figure 4.2 makes the contrast.

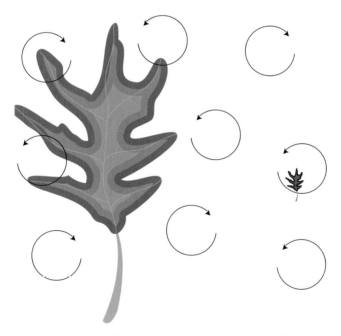

FIGURE 4.2. How an object, here a leaf, experiences vortices depends on its size relative to the size of typical vortices. If you're large and they're small, you feel them as generalized turbulence. If you're small and they're large, you know little more than you're moving in circles—and that only if you have the necessary sensory capability.

What about a leaf in the kind of minimally moving air that surrounds it most of the time? In just such low-speed air movement, the balance between laminar and turbulent flow becomes uncertain while the distinction becomes particularly relevant. A yardstick for judging whether the leaf experiences laminar or turbulent air is the Kolmogorov length scale, named for A. N. Kolmogorov, the mathematician who derived it in 1941. It gives the size of the smallest eddy under various circumstances. That size depends on how viscous the fluid is—higher viscosity means that the smallest eddies won't be as small. And it depends on how rapidly energy is being dissipated—energy rattling around means that smaller eddies will exist. Keeping eddies spinning takes energy—viscosity extracts its price, and the smaller (and thus more numerous) the eddies, the more energy they eat up.* For an individual leaf, minimal airflow will most likely be laminar, even if a large plume of smoke indicates turbulence.

We immediately ask, or ought to ask, if bulk flow is so great, why should diffusive transport of the CO_2 in the air to a leaf's surface still matter at all? Chapter 3 hinted at the answer when noting that diffusion is a factor in olfaction only very close to our actual nasal surface. To see why diffusion retains relevance if any decent number of the critical molecules is to reach the actual surface, we have to describe a peculiarity of the way fluids flow, a critical consequence of viscosity.

When a fluid flows across a solid surface—any fluid, any speed, any surface—the speed of flow at the surface is zero. This is the so-called no-slip condition. Its existence (or nonexistence) occasioned hot debate back in the mid-nineteenth century, but it fits so perfectly with both theory and measurement that it has long outlived any controversy. So where does the real flow begin? Just beyond some molecularly thick layer on the surface, which is to say beyond a negligibly thin layer. But this real flow begins as very slow flow near the surface, gradually speeding up at greater

---

*The formula for the Kolmogorov length scale is simple enough, although putting numbers into it isn't always quite so easy:

$$l_K - \left(\frac{\mu^3}{\rho^3 \varepsilon} \right)^{1/4}.$$

μ and ρ (lowercase rho) are viscosity and density, respectively, while ε (lowercase epsilon) is the rate of energy dissipation per unit mass. This last requires measurement of three components of velocity and plugging them into an equation for kinetic energy. I don't think it's been done for these minimal winds in the vicinity of foliage, but I expect the length will be well above the length or width of typical leaves.

distances until it eventually reaches (put strictly, it approaches) the full speed of stream, bloodstream, or wind—as in figure 4.3.

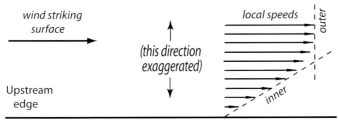

FIGURE 4.3. A fluid moving across a solid surface behaves quite unlike one solid moving across another. Lengths of arrows at the right show relative flow speeds. At the actual surface, the fluid's speed is zero—or that of the surface—so a region develops in which speed increases with distance from the surface. The outward increase in speed within the inner part of the region is linear, while across the outer part, the speed gradually approaches ("asymptotically") the speed of the fluid far from the surface. That gradient region, generally very thin, gets thicker the further along the surface the fluid moves.

What demands this gradual approach to full flow is the fact that fluids (even gases) have viscosity, and so they resist shear—sliding across themselves. The no-slip condition has an inescapable corollary. Everywhere that a fluid flows across a surface, it forms a speed gradient, from zero speed at the surface to full flow some finite distance outward. So while the zero-velocity layer may be negligibly thick, speed will change within a significantly thick gradient region. The higher the overall flow speed, though, the thinner that gradient region and the shorter the distance left for diffusion to provide transportation.

In effect, viscosity amounts to a kind of internal friction in fluids. Ultimately, it's why you need a heart to pump blood around. If viscosity, effective friction, were zero, then blood, once started, would continue to circulate the way the moon continues to circle the earth. But viscosity enforces that no-slip condition and the resulting variation in flow speed across every blood vessel. Variation in flow speed means that fluid has to slide across itself to flow along. Fluid doesn't like to do that, so it must be coaxed with energy, and so a pump must put energy into the bloodstream to keep it moving. The speed gradient, again, occurs whether the flow is laminar or turbulent and everywhere that a fluid flows across a surface, whether blood vessel wall or leaf. Even when turbulent flow produces a generally turbulent speed gradient, there's still a laminar region right near the surface.

The no-slip condition makes trouble—if subtle trouble—in our everyday lives. One swipe of a dishcloth works as well as lot of flowing hot water. The dishcloth contacts the dish, sweeping away surface mess; the flowing water doesn't make such effective contact. Hot water works better than cold, not just because more substances dissolve in hot water, but because hot water has a much lower viscosity; that means more fluid moves close to the dish. Dust accumulates on the surfaces of fan blades; the low flow right near a blade's surface isn't enough to dislodge it.

One particular peculiarity of this speed gradient bears strongly on the design of leaves. Imagine flow going across a flat plate, starting at an edge. Near the edge the speed gradient is steep—full speed flow occurs not too far out from the surface. But as the fluid passes along the surface, it becomes increasingly aware of its own viscosity, and its speed gradient, as, again, in figure 4.3, gets ever gentler. As a result, full flow occurs ever further away from the surface. That in turn means that the effectiveness of wind in bringing CO_2 within diffusion range becomes ever poorer. Since it involves cross-flow transport, turbulence might help bring CO_2 closer, but as the Kolmogorov length scale suggests, in very low winds eddies will be large, and flow across a leaf should be laminar.

The problem can be solved by making narrow leaves, leaves over which air needn't ever flow very far. That, though, interacts with the ever-critical business of light interception, with its nod going to broader leaves. We see a world of compromises, a world in which no one design may be absolutely optimal, a world in which the best trade-off point gets seriously situation specific. Neither CO_2 concentration nor the intensity of full sunlight varies much from place to place, but just about everything else does. Broad leaves may be common now, but they arrived fairly late in the evolution of terrestrial plants. On that basis alone, we suspect that designing effective ones presents a real challenge for natural selection.

At this point a number or two might be informative. Consider a very low wind, one that might be low enough to make trouble but still high enough to occur in nature. Measurements I made years ago suggest 0.1 meters per second, about 4 inches per second or a quarter of a mile per hour, as a wind that might characterize a still period during a generally still day. You can't feel a wind this low, but an appropriate instrument can easily detect it. And consider a leaf that's about 5 centimeters, approximately 2 inches, across. How thick will the gradient region be when the wind reaches the downstream edge of the leaf?* We can't specify an exact distance, since the local wind speed approaches the real, undisturbed wind ever more slowly with distance from the surface. So (arbitrarily and conveniently) we'll look at the thickness calculated outward from the leaf's surface to where the wind has reached 90 percent of its full strength. The thickness comes out to about 8 millimeters, a little over a quarter of an inch.

In our world that's not much. But it's dauntingly remote from the vantage point of diffusion, and diffusion must take over where bulk flow fails. At least the air in the region we're talking about isn't motionless, but just moves more slowly than our already slow 0.1 meters per second. Still, we're looking at a potential problem. Many leaves have unbroken widths greater than 5 centimeters. Most of these, though, live in well-shaded places, where the limited light might enforce a lifestyle leisurely enough to make a lower CO_2 supply tolerable. But as we'll see in subsequent chapters, CO_2 access isn't the only problem raised by making broad leaves broad.

Bottom line: bulk flow of one sort or another takes care of long-distance transport, transport on our perceptual scale, with diffusion taking over on the micro scale, whether near the surface of a leaf or near our nasal epithelia.

A Tool for Trade-Offs

Diffusion versus bulk flow—we're looking at the interplay of two fundamentally different mechanisms that transport gas molecules between air

---

*The operative formula for what's called boundary layer thickness for laminar flow to the 90 percent point is

$$\delta = 3.5\sqrt{\frac{x\mu}{v\rho}},$$

where δ (lowercase delta) is layer thickness, x is distance downstream from upstream edge, v is distant flow speed, μ is the air's viscosity, and ρ is air density. The commoner formula uses the 99 percent point, which is misleadingly thick for our purposes—in it the 3.5 is replaced by 5.

and leaf. Which one matters where and when? We might guess that that depends on factors such as wind speed, leaf size, and the behavior of the particular kind of gas molecule. Engineers, particularly those working in fluid mechanics, have given us a terrific tool to help in making such judgments. The tool is simple but also crude: simple enough so even a nonscientist (or a biologist) needn't be intimidated, but crude enough so it provides only rough guidance and provokes sneers from purists. Physicists rarely use it and few mathematicians need it, so it passes under the radars of their undergraduate math and science courses.* (Kolmogorov, an exceptional mathematician, derived his eddy size scale with it.)

The basic idea is laughably ordinary: divide the value of one factor by the value of the other factor. What you get are simple ratios, but these ratios share one formal and oddly relevant feature. Done by the book, the two factors should have the same basic variables. That makes them dimensionless. Dividing distance by distance, for instance, gives a pure number: it's three times as far from my home to the drugstore as it is from home to my office. Keeping the ratios dimensionless keeps them unitless in the bargain: three times, not 3 kilometers or 3 miles. It's strange, even jarring, to think that a number that says something about the real world can lack units, but I've just conjured one up without even breaking a sweat. What fraction of your life do you spend sleeping? What's the annual yield of an account offered by a bank? Two others. Stranger, the answers to such questions, those dimensionless ratios, come out precisely the same whether you enter the measurements in hours or seconds, in dollars or euros—as long as you start by putting both numbers in the same units.

In the same way, we can create a ratio of the rate of transport by bulk fluid flow to the rate of transport by diffusion. If the value is well above 1, then the system is flow dominated; if the value comes out well below 1, then the system is diffusion dominated. This particular ratio is called the Péclet number, in honor of Jean Claude Eugène Péclet (1793–1857), a French physicist. It's simple. Multiply the relevant length of the system by the speed of flow and divide by the diffusion coefficient for whatever substance you have in mind.† Again, you are enjoined (on pain of self-

*The formal name for the business is dimensional analysis.

† As an equation, where $Pé$ is the Péclet number, v is flow speed, x is distance, and D is the diffusion coefficient,

deception) to put all the factors in the same system of units. Most often (but not necessarily) we use meters for distance, seconds for time, kilograms for mass, newtons for force, joules for energy, watts for power—so-called SI units.

Back to the leaf and its access to CO_2 with a calculation similar to one made earlier. A representative distance from an edge for a broad leaf might be around 20 millimeters; a typical minimal flow speed should be around 0.1 meters per second. That means the thickness of the speed gradient is around 6 millimeters. Multiply that distance and speed (converting 6 millimeters to 0.006 meters) and then divide by the diffusion coefficient for CO_2 in air. You get a value of around 40—no use carrying out a more precise calculation for this kind of a ballpark rule. Bulk flow, here flow of air, clearly dominates this system. Put in practical terms, in describing what's happening, you can't ignore the influence of wind, even wind too slow to feel. For the Péclet number to drop to 1, the wind would have to be forty times lower yet, lower than probably ever occurs in nature for more than a fraction of a second (during direction reversals, perhaps). Right at the leaf's surface, diffusion has to matter, but it cannot by any means be the whole story, even as a crude approximation—contrary to the assumption, by the way, in some textbook accounts of gas exchange in leaves.

That value of the Péclet number says a bit more. For one thing, the value tells us that changes in wind speed should have little practical impact on a leaf's access to CO_2. That's because even the very low wind we assumed pushes the number far above 1. For another, a value so far above 1 suggests that the system does more with flow than it needs to. If we were talking about a circulatory system, we'd immediately wonder if we'd done our measurements or calculation correctly. Thus the value for blood flow and oxygen diffusion in capillaries ought to be about 1. Otherwise, the heart would be pumping an unnecessary volume of blood (if the Péclet number were high) or capillaries would be unnecessarily small, making

---

$$Pé = \frac{vx}{D}.$$

Only a few diffusion coefficients are of interest to us. Rough values (they vary with temperature and other factors) for diffusion in air are, for CO_2, 0.15; for O_2, 0.18; and for H_2O, 0.24 centimeters squared per second. In water, the value for CO_2 is 14×10^{-6}; for O_2, 20×10^{-6} centimeters squared per second. The units commonly cited are anachronistic. To convert to proper SI, meters squared per second, divide by 10,000.

the heart produce too much pressure (if the Péclet number were too low). In fact, the radius of capillaries (3 micrometers), the speed of flow (0.7 millimeters per second), and the diffusion coefficient of oxygen combine to give a value of 1.2, remarkably close to that ideal of 1. For leaves, though, the question may be moot—wind comes at no charge, so profligate use exacts no penalty in cost.

Further along, we'll have repeated recourse to the kind of rough-and-ready tool afforded by dimensionless numbers. They rarely give a full or precise answer to a question, but they provide enormous help in keeping us from pursuing will o'the wisps into cul-de-sacs, as my late graduate adviser was wont to say. It's all too tempting (in my humbling experience, at least) to chase a hypothesis that, however exciting it sounds qualitatively, turns out to be beneath notice in any quantitatively significant sense. The practicing scientist can never have too many crap detectors.

Pushing Péclet Further

What else might depend on the particular mix of diffusion and forced flow? Some stomata should see speedier airflow across their outside opening than should others—particularly those nearer the edges of leaves, where speed gradients will be steeper. And stomata might work like chimneys, where wind across the top sucks air and smoke more rapidly upward. In particular, the stomata on plants that live in moist places often have slight chimneylike extensions on the outer rims of their opening. Chimneys might induce bulk airflow—wind—in the spongy mesophyll of leaves, possibly taking some of the burden of transport from diffusion. That would reduce the need for a concentration difference and therefore expose the photosynthetic cells to higher levels of CO_2. Quite a few years ago, I thought the notion sufficiently intriguing for an experimental investigation. Leaves just had to be doing something so neat, but the literature said nothing about wind within leaves. After months of work, a series of technical obstacles persuaded me to shift attention elsewhere. Perhaps that was just as well—I wasn't acquainted with Monsieur Péclet at the time.

How much wind would it take to make a significant difference? Enough, one supposes, for the Péclet number to approach 1. The relevant distance is about 0.2 millimeter, about as far as diffusion ever has to move

CO_2 from a stomate to a cell within a leaf. That means an internal wind of around 7.5 millimeters per second would be needed. For an atmospheric wind, that's no big deal, but internal winds that strong are highly unlikely. First, routine external winds aren't high enough. Second, the overall aperture area of the stomata of ordinary leaves isn't great enough to generate anything like the pressures needed to drive such internal wind through small, tortuous passages—recall figure 3.3. Sure, there has to be some internal bulk airflow, but a calculation of Péclet number argues that it's unlikely to be consequential in any functional sense. A quick back-of-the-envelope check might have saved me a lot of trouble—this was one of the humbling experiences alluded to earlier. Or, better, calculations might have directed my attention to particular leaves that had the right structure and lived in the right places to make use of the scheme. (See, I haven't quite given up. Pet hypotheses die hard; we scientists are only human.)

Calculations of Péclet numbers tell another inside story that's both more general and more credible. The cells that make up leaves and most of the other parts of most plants are large by comparison with our own cells, roughly 50 or 100 micrometers rather than 5 or 10 micrometers across (fig. 4.4). Diffusion as a transport mode should work 10 squared,

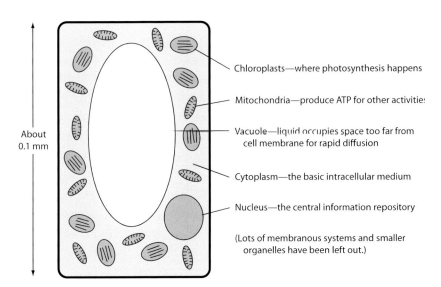

About
0.1 mm

Chloroplasts—where photosynthesis happens

Mitochondria—produce ATP for other activities

Vacuole—liquid occupies space too far from cell membrane for rapid diffusion

Cytoplasm—the basic intracellular medium

Nucleus—the central information repository

(Lots of membranous systems and smaller organelles have been left out.)

FIGURE 4.4. Another severely idealized drawing, this one of a typical plant cell. As noted, lots has been left out, and "typical" sweeps a great range of variation under the rug.

or 100 times, less effectively within such large cells. And that would enforce a far more leisurely pace of metabolic life. But unlike animal cells, in which predominant agency of internal transport is diffusion, plant cells depend less exclusively on diffusion. Plant cells supplement it (again, no biological system ever fully replaces it) with internal circulation. The contents of plant cells typically move around (called cytoplasmic streaming or cyclosis), driven by a protein motor with some biochemical and functional similarities to our own muscle. Typical flow speeds are around 5 micrometers per second. That may sound slow to us, but it's enough to move fluid fairly fast relative to the size of cells. A subcellular item can make an entire circuit of such a plant cell in about the same time as a bit of blood takes to make a full circuit of the circulatory system of an active human.

Are the large cell size and the presence of internal circulation related, as we (at least I) would like to assert? For a cell radius of 50 micrometers (the distance needed to get to the middle) and water as the medium, the Péclet number for CO_2 comes out to 0.15. So even with that flow, diffusion provides the main means of transport—much faster flow would be needed for flow to help. Or, assuming a Péclet number of 1, the effective cell radius could be all of a third of a millimeter. Perhaps plant cells could be far larger still. But I think that misleads us. As a small molecule, CO_2 diffuses readily. These cells have to move protein molecules around, molecules hundreds of times larger.

What if we turn things around and assume that a plant cell does something critical with a Péclet number of about 1? From that, the speed of cyclosis, and the size of the plant cell, we calculate a much lower diffusion coefficient than that for CO_2. And from this diffusion coefficient, it's no big stretch to figure out a molecular size—or macromolecular size, since the larger the molecule, the lower the diffusion coefficient.* That Péclet number of 1 corresponds to diffusion of molecules that are larger than dissolved sugars and amino acids but smaller than most proteins. The calculation suggests—to me at least—that plant cells rely predominantly

*Diffusion coefficients vary inversely with the square roots of molecular weights, at least roughly, since the shapes of molecules (and other things) differ. That once provided an important tool in figuring molecular weights. (For gases at constant pressure, density and molecular weight vary in just the same way.)

on flow for transport of proteins and other really large molecules. By contrast, middle-sized and small molecules move as much or more by diffusion, speedy enough for them to do the jobs needed to maintain a decent pace of life.

Two final points. First, look at all the mileage we got from a simple ratio—trivial mathematically, but so remarkably revealing of why the living world is as it is. One can apply the Péclet number to many more situations, something I've done elsewhere.[2] And second, we're tickling the edge of the larger issue of size and scaling with this discussion of the relationship between cell size and the need for flow to augment diffusion. How the particular physical factors relevant to effective design depend on an entity's size matters again and again when looking at how organisms are organized. So explanations based on size—how phenomena scale up or down—will appear repeatedly in the chapters to come.

TO REVIEW, the leaf's business, photosynthesis, needs three main inputs: light, for energy, and carbon dioxide plus water, for starting materials. The first of these full sunlight provides in generous measure. CO_2, at just under 0.04 percent of the atmosphere's composition, is scarce, and our concern with how to access it generated all that talk in the past two chapters about diffusion and flow. Should we worry about water? Photosynthesis requires roughly the same number of molecules of water as it does of CO_2. At first glance, supplying it looks like a nonproblem. If CO_2 can be fed in as a gas, surely water can be fed into the plant system in the same way. CO_2 may be scarce, even as we humans push up its atmospheric fraction, but air holds lots of water vapor. Just how much depends on temperature and relative humidity.

So—a few values. It's easiest to give water content (mass of water vapor relative to mass of the whole mix) for 100 percent relative humidity, that is, saturation.* Then we just have to adjust downward as needed, for instance multiplying by 0.25 for 25 percent relative humidity. At the freezing point, the maximum water content is 0.38 percent of the mix; at 20°C (68°F), it's 1.47 percent; at 40°C (104°F), it's 4.9 percent.[1] Put another way, you'd have to have rather cold, dry air for it to have as little water as it has CO_2—for instance 0°C (32°F) with 8 percent relative humidity. Even better for photosynthesis, water vapor's diffusion coefficient is about a third higher than that of CO_2. Water supply should present no problem except perhaps in a cold desert.

Why So Much Water?

Paradoxically, getting enough water can present a very real problem—no surprise to anyone who waters houseplants,

---

*Water content, as mass of water relative to overall mass of air in any arbitrary volume, is termed absolute humidity. It's all too rarely mentioned, so we don't ordinarily hear or else easily forget how much water vapor air may contain.

looks skyward for rain, or irrigates a crop. Plants, like animals, contain as much or more water than all their other constituents put together. Thus growth requires an input of water beyond what's needed for photosynthesis itself. Still, that amounts to a once-and-for-all requirement, a minor business except when an initially dry seed must absorb water to germinate. Sure, most of the weight of a leaf represents water, but the leaf needs to be made only once, or at most once per year.

We humans need lots of water—where does ours go? A lot goes out in our exhalations, which come out close to body temperature and 100 percent humidity. So you might inhale air with 1 percent water vapor and exhale air with about 3.5 percent. In nothing like this sense does a leaf need to breathe. Even at night, when it uses oxygen to respire the same way we do all the time, its flatness makes direct diffusional and wind-driven access entirely practical. A lot of water leaves through our kidneys, since they can dump wastes only when they're well diluted with the stuff. Leaves absorb or synthesize only what they need, so their excretory needs are trivial. Some water evaporates from our skin, but mainly when cooling demands that we perspire. Many leaves engage in a bit of evaporative cooling, but it's no absolute requirement for them either.

One might regard a leaf's profligate demand for water as evidence of outrageously bad design. For some plants, 97 percent of the water they absorb—almost entirely as liquid entering their roots—just . . . well, it just disappears into thin air, almost all through their leaves. Rather than diffusing in through the stomata as does CO_2, water vapor diffuses out, often in prodigious amounts. Trees manage to suck the stuff out of the soil, even out of fairly dry soil, and they then lift it up to their leaves. Then they simply allow almost all that water to evaporate out into the atmosphere. We're talking about big-time hydraulic heroics: a large tree may lift as much as 110 gallons, or 500 liters, every day. Hard work, lifting all that water so high, work in precisely the sense used in physics— work demanding an input of energy. Still, the tree does the work slowly and steadily. That 30-meter (100-foot) tree, lifting water for 14 hours a day (trees mostly take the nights off), must work at a minimum rate of 3 watts.* (We'll ask about the source of energy for that work later.)

*Power is work per unit time or rate of working, calculated here as mass lifted (in kilograms; a liter of water weighs 1 kilogram) times height of lift (in meters) times

For comparison, a fit human can lift steadily at an output of about 100 watts,[2] although the nonfanatic finds that far from pleasant.* Relative to our weight, we can work harder at lifting than can trees. Still, one rarely thinks of plants doing mechanical (as opposed to chemical) work.

Yes, as mentioned, some leaves sometimes cool themselves by evaporating water, an activity to which we'll return in chapter 8. But even when a leaf's temperature puts it at no hazard of thermal injury or even of reduced photosynthetic activity, it emits water vapor. The classic explanation for this huge loss of water carries the strong odor of a maladaptive disadvantage for the leaf, and of a desperate default for the plant physiologist. Put bluntly, that evaporative water loss—or transpiration—was long regarded as an unavoidable evil. Since we know of many other problems that evolution can't seem to solve, we needn't dismiss the rationalization on that score alone.

Because the leaf is open to influx of CO_2, either diffusively or through flow, it's also open to efflux of water—that's the heart of the problem. To be used, gaseous CO_2 has to go into solution on wet surfaces within the leaf; therefore water will be lost from those same surfaces. Only if the relative humidity is 100 percent will water not be lost. If the leaf's temperature is above that of the surrounding air, then water can be lost even at that humidity. We might imagine a leaf that keeps its temperature below ambient and thus vaporizes less water, but at least during the day that's hard to do by any means other than evaporative cooling itself. Moreover, the higher diffusion coefficient of H_2O than of CO_2 means that the molecular rate of water loss will exceed the rate of carbon dioxide gain, even assuming that the same concentration gradient drove the two diffusive streams. In fact, the gradient for H_2O will be far greater and thus a more potent driver.

Let's go about this business of water loss by leaves step by step. We

---

earth's gravitational acceleration (9.8 meters per second squared) divided by time (in seconds) to get watts in SI units. A watt is 0.239 calories per second or 0.860 kilocalories per hour, the units on some exercise machines.

*The figures you note when exercising on a calibrated treadmill, rowing machine, or similar substitute for productive work represent an estimate of power input, that is, metabolic rate, not output. To get power output, you have to divide by 4 or 5. The other 75 or 80 percent comes out as heat, obeying the second law of thermodynamics and then some. That figure of 100 watts for sustained power output has a fascinating 250-year history, which I described in an earlier book.

might start with the most naïve presumption—where I started, to be honest. Net CO_2 diffusion goes inward, net H_2O diffusion goes outward, with a one-to-one exchange of molecules, or 1 gram of CO_2 inward for each gram of O_2 outward. No, that isn't quite right—we can't move so blithely from molecules to mass. CO_2 molecules are over twice as heavy as H_2O molecules, so a one-to-one molecular exchange would have 2.4 times as much CO_2 going in as H_2O coming out, gram for gram (or, since that's a dimensionless ratio, megaton for megaton, same thing). Better, but not really right. As just mentioned, H_2O, being a smaller molecule, has a higher diffusion coefficient than CO_2 and will move more readily, which eats up most of that gas's advantage, reducing it to 1.6 times more CO_2 than H_2O. A little better, but not by much.

We're still being pretty naïve. Big-time trouble comes from the concentration differences, which, after all, determine net diffusion quite as much as do the diffusion coefficients. CO_2, for all our contemporary complaining, remains a rare gas—again, our leaf's ancestors can be blamed for depleting the resource. So even after serious human augmentation, the maximum concentration difference can't exceed about 0.038 percent. And again, water vapor would be that scarce only in a very dry, cold atmosphere. Inside the leaf, behind the stomata, the air will be saturated with water vapor; on a warm, clear day the level outside will be far less, so the difference might be around 3 or 4 percent. That's a hundred times greater than the CO_2 difference, meaning that only a tiny bit of carbon dioxide will be gained relative to water lost—in our example, that 1.6 gets divided by 100, giving not 1.6 but 0.016 grams of CO_2 going in per gram of H_2O going out. An interest on the liquid investment of 1.6%, we might say.

Nor are we quite done with our discounting. A leaf can't run the CO_2 concentration inside down to zero. Basic thermodynamics insists that proper processes, living or not, can't work that effectively. So the residual CO_2 level behind the stomata will be higher, reducing the concentration difference that drives inward diffusion. A typical value for a leaf on a tree might be 0.02 percent, leaving a concentration difference from outside (0.038 percent, again) of less than 0.02 percent. That reduces the expected ratio of CO_2 going in to water coming out from 0.016 to around 0.008—less than 1 percent. Put another way, for every gram of CO_2 the leaf uses, it should lose something on the order of 125 grams of H_2O. Maybe that's simply the cost of living in our particular world.

No wonder that irrigation is such an ancient and widespread practice. No wonder that irrigation alone accounts for 70 to 80 percent (depending on whose estimate we adopt) of the water consumed by our species. Water consumption takes on especial contemporary urgency, not just because we've been increasing in numbers so prolifically in the past decades, but because we're becoming serious about generating biofuels. That will require that we bring more marginal land into agricultural use, and "marginal" tends to mean dry. Worse, no source of energy comes close to biofuels in its requirement for water. A recent paper gives some alarming numbers. Extracting and refining oil takes less than 200 liters of water per megawatt hour of energy made available. Nuclear power uses around 1000 liters per megawatt hour without external water cooling or about 150,000 with water cooling. But making corn-based ethanol demands about 5 million liters per megawatt hour, and making soybean-based biodiesel needs around four times that amount.[3] At least, thank heaven, water can be managed as a renewable resource.

The most straightforward fix for low water-use efficiency is an increase in ambient CO_2 concentration, just what we're now doing with the earth's atmosphere. Increase CO_2, and you ordinarily both increase a plant's productivity and decrease its water use. In greenhouses CO_2 can be (and sometimes is) increased to good effect. If those were the only effects of elevated CO_2, we'd be in clover, so to speak. The bad news: every other effect is bad, more than offsetting any gain on anything other than a local and temporary basis. Another fix consists of raising the humidity. That's a little hard to arrange outdoors without diverting at least as much water as might be saved. Still, it may be an advantage for raising plants in dense groups, whether the groups are intra- or interspecific. They may compete for light and other resources, but their overall transpiration will be less than the sum of what each would produce in isolation.

Tricks and Trade-Offs for Effective Water Use

Not that plants—some plants—don't have evasions up their sleeves, schemes to increase carbon capture relative to water loss. That quantity, water use efficiency (often going by the acronym WUE in the trade), isn't some unalterable biogeochemical constant or even a physically determined given. First, though, a note about the term *water use efficiency*.

While the underlying idea combines importance and simplicity, in practice it's a mess, because it means different things to different communities. WUE can be expressed as the number of molecules of CO_2 coming in versus the number of H_2O's going out, or the mass of one versus the mass of the other—again, not the same thing. Either computation can be an instantaneous value, a daytime average, a day-plus-night average, or a growing-season average. Instead of CO_2 absorbed, the ratio may substitute carbon, C, chemically fixed by a plant. For the crop scientists, it commonly includes evaporation from the soil in their accounting of water use, that is, water that never entered the plant at all. Or it may add runoff to evaporation to give even more agriculturally relevant data.[*][4]

For our more limited and admittedly simplistic viewpoint, that of the instantaneous world of the individual leaf, I'll use mass of CO_2 divided by mass of H_2O for an actively photosynthesizing plant.[†] A typical value of around 0.025 gives some idea of how much water is lost relative to CO_2 gained—forty times as much (1 divided by 0.025 equals 40). That's better than our previous estimate of 125 times but still pretty awful, so the basic question remains the same. What can be done to improve WUE when water is an uncertain or expensive resource? That's a question that must have come up, even if only vaguely recognized, when humans first had to choose which seeds they should plant, then where to plant those seeds, and then how much water they should laboriously lift from pond or river to the growing crops.

What's behind the difference between our estimate of a water use 125 times the CO_2 gain (WUE = 0.008) and this typical value of only 40 times as much (WUE = 0.025)? It looks as if plants have somehow mitigated their unfavorable physical legacy. Our simple model fails, which means something interesting must be going on. In my experience, models are often

---

[*] Not only that, it stretches the meaning of efficiency, which ordinarily has the same variable in numerator and denominator—for instance, in the case of motors, power output over power input. The best that can be said about this "efficiency" is that it's properly dimensionless, with the same units top and bottom.

[†] As pointed out, CO_2, with a molecular weight of 44, has heavier molecules than does H_2O, whose molecular weight is 18. Thus you can shift WUE from a mass basis to numbers of molecules by multiplying by $\frac{18}{44}$, or 0.41—or shift from molecules to mass by multiplying by 2.44. Yet another set of measures are based on the reciprocals of all versions of WUE, generally termed transpiration ratios.

as revealing—or at least prescient—when they fail as when they succeed in matching reality. So what might be going on, what manner of tricks might come into play? Three distinct devices stand out: two manipulate metabolic chemistry, the third a physical scheme.

C_3 **versus C_4 photosynthesis.** In the classic description of the molecular events in photosynthesis, CO_2 first combines with a five-carbon compound to form a pair of three-carbon intermediates. All these compounds have names, which matter so little for present purposes that I'll not divulge them. That follows my first principle of writing: "Explain, dammit, don't just mention." Anyway, there's always *Wikipedia* for the truly inquisitive. The scheme, which ultimately regenerates the five-carbon compound, bears the name of its discoverer, Melvin Calvin: the Calvin cycle. (As it's also a combination of the first names of twins in my elementary school class, I never need to look up the name.) We use an analogous circular scheme, the Krebs cycle (after Sir Hans Krebs), as we break sugars down to CO_2 to supply chemical energy for our bodily activities. Nature seems to prefer to handle one-carbon compounds as little as possible—maybe these have too great a penchant for diffusing away if left even briefly unbound.

An alternative to the Calvin cycle, which has been known since the 1940s, was described in the 1960s. Instead of initially forming C_3 compounds, this one combines CO_2 with different intermediates to form one or another C_4 compounds. So we now distinguish between C_3 photosynthesis and C_4 photosynthesis, sometimes called the Hatch-Slack cycle after its discoverers. Not only was the C_3 scheme worked out earlier, it seems to be the ur-form, the default arrangement. From C_3 precursors, the C_4 scheme has evolved in a surprisingly large number of lineages—in some way its basic elements must be universally represented in the genetic material of green plants.

C_4 photosynthesis occurs in only about 1 percent of vascular plant species (loosely, all terrestrial plants except mosses), but these C_4 species make up about 5 percent of the earth's mass of living plants, and they contribute a still higher fraction of its photosynthetic production. C_4 photosynthesis is commonest among grasses, desert shrubs, and a disproportionately large number of the weeds we consider noxious. Among our major crops, corn (maize) uses C_4 photosynthesis, while wheat and rice are C_3 plants. Deciduous and coniferous trees—and thus our paradigmatic leaf—are C_3 plants.

Why bring up this arcane item of metabolic biochemistry in the present context? Simply because the two variants differ, at least on average, in WUE. The C_4 plants fare significantly better, losing about half as much water for each unit of CO_2 they acquire. They're also more productive, in part because their photosynthetic gear can use brighter light—C_3 plants max out well below the light levels of direct sunlight. C_4 plants cannot just tolerate stronger light—they increase their productivity when in light three or more times more intense than intensities that fully supply C_3 plants. They also hit maximum productivity at higher temperatures, typically around 40° rather than around 30°. As Thomas Sinclair points out, if you want to make biofuels, C_4 plants are head and shoulders (twigs and branches?) ahead.[5]

C_4 plants derive their advantage in WUE from a more effective CO_2 trap. They can drive CO_2 levels in their internal air spaces from the atmospheric 0.038 percent to 0.009 percent—rather than the 0.021 percent typical of C_3 plants. That gives them a greater concentration difference with which to drive net inward diffusion.

What's the downside? After all, if there weren't some compensatory disability, evolution, rewarding superiority, would have bequeathed us far more C_4 forms than the present 1 percent of plant species. Photosynthesis takes energy in the form of sunlight, and C_4 photosynthesis takes more energy than does C_3 photosynthesis. So while these peculiar plants not only have the ability to use higher intensities of light—up to that of an overhead sun—their less light-efficient photosynthesis requires higher intensities. Still, the sun is certainly bright enough for them and often needlessly bright for the C_3-ers—again, what's the dark side of the device? The obvious problem is that full direct sunlight can't be assumed. If more light is needed, then the effective day length is less. For a plant, effective day length is the period during which photosynthetic production exceeds respiratory consumption, the period when plants are net consumers of CO_2 and net producers of oxygen. Moreover, the sun may be obscured by clouds or—big or—a leaf may be shadowed by other parts of the plant itself or by its neighbors.

Perhaps that's why ordinary trees don't use C_4 photosynthesis, assuming (not unlikely) that they could have evolved the pathway. What might be useful for trees but doesn't seem to occur would be to make C_4 leaves near the top and C_3 leaves as their shaded brethren below. Two kinds of leaves on the same tree and therefore with the same genetic makeup? Since sunlit and shaded leaves often differ markedly in morphology (chapter 8 will fuss about that), this scenario is not as far-fetched as it sounds. But trees don't seem to follow it.

For me, the C_3 versus C_4 distinction couldn't be closer to home. Each fall I renew my lawn with seed, lime, and fertilizer. I can pick a C_3 grass, fescue (*Festuca*), or a C_4 one, Bermuda grass (*Cynodon*)—the store carries both. The yard is a mixed bag, with some shady parts and one south-facing sunny slope. If I plant fescue, it does better in the shade and stays nicely green in the winter, but then goes brown over the summer, beginning with the slope. If I plant the more heat-, light-, and drought-resistant Bermuda grass, it's brown in the winter and grows poorly in the shade. Usually I choose fescue, because our windows look out on the shady portions, while only the neighbors see the sunny slope. Also, it's cheaper.

CAM metabolism. This is yet another of nature's tricks, a particularly lovely one. At night, temperatures near the earth's surface drop—there's no input from the sun and often a lot of reradiation to a cold sky. That pushes the relative humidity up, even with, as explained earlier, the same absolute concentration of water vapor in the air. So if stomata open at night, a lot less water will be lost. That's the basis of so-called CAM photosynthesis, a variant (although discovered earlier) of the C_4 game. CAM stands for crassulacean acid metabolism, an awkward mouthful, named for the plant family Crassulaceae, in which the mechanism was first recognized. The most familiar plants in that family are pineapples.

CAM plants keep their stomata closed during the day, opening them only at night. While that greatly reduces water loss, it provides access to CO_2 at precisely the worst time; after all, no sunlight means no photosynthesis. At night CAM plants convert CO_2 to the same four carbon compounds as do the C_4 plants. But instead of going further in the Hatch-Slack cycle, they store up the CO_2 in the form of organic acids. These then get fed back into the cycle for the rest of their processing after the sun comes up. C_4 plants may have twice the WUE of C_3 plants; CAM C_4's manage another twofold increase.

Like the C_4 scheme, CAM photosynthesis has convergently evolved in a large number of lineages. All cacti use it, as well as a lot of orchids, bromeliads, euphorbs, aloes, yuccas, and others. It, too, has a downside: the amount of organic acid that can be stored without making trouble. Storing anything in soluble form can run into difficulties from either chemical toxicity or simple physical concentration—this last mainly by upsetting the osmotic condition of cells, about which more shortly. Thus, while its greater WUE might allow life in otherwise marginal habitats, CAM photosynthesis doesn't translate into especially high photosynthetic productivity.

Stomatal adjustment. This is the physical scheme. Leaves control passage in and out by adjusting the area of their stomata—from wide open, when they provide only minimal resistance to the overall path of diffusion, to fully closed, when diffusive exchange approaches zero. One might expect—at least I did until I read otherwise—that the change in resistance would affect CO_2 and water vapor equally. In fact, changes in stomatal aperture affect the diffusion of water more than they do the diffusion of CO_2. That's not because water, with lighter molecules, diffuses more readily. Rather, it's because fewer elements resist diffusion of water than resist diffusion of CO_2.

The paths for gas exchange between leaf and environment can be viewed as resistances in series, as in figure 5.1. For water vapor, three main elements form the series: the resistance of the internal air spaces, the resistance to passage through the stomata, and the resistance of the semistagnant air on the outer surface of the leaf. Getting from cells to internal air presents no problem, inasmuch as water evaporates readily from the wet internal surfaces, and the wet internal surfaces have lots of area twenty to forty times that of the outer leaf surface. That's why leaves have internal relative humidities close to 100 percent.

By contrast, CO_2 diffusion faces a longer series of resistances. In addition to water's three main elements, it meets resistance as it diffuses across cell walls and membranes, through the cytoplasm of the photosynthetic cells, and then into their chloroplasts—the bodies that do the business. That these additional resistances involve CO_2 diffusing within a liquid rather than as a gas in air makes no fundamental difference. Adding additional elements means that adjusting any one element exerts less overall effect. Imagine a series of three resistive elements (as for water

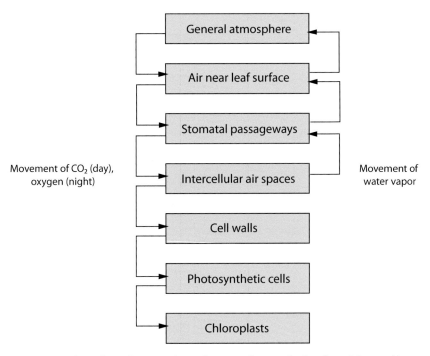

Movement of CO_2 (day), oxygen (night)

Movement of water vapor

FIGURE 5.1. The pathway for gas exchange between photosynthetic cells and the outside atmosphere. Losing water, moving upward on the right, faces fewer barriers than does acquiring CO_2, going downward on the left. Bear in mind that if the process is to proceed, some positive concentration gradient (or its equivalent for gas-liquid transitions) must be maintained at each stage.

vapor), each with a resistance of one unit. Raising one element to two increases overall resistance from three to four, a 33 percent increase. Then imagine a series of six elements (as for CO_2), each with an initial resistance of one unit. Raising one element to two increases overall resistance from six to seven, an increase of only 17 percent.

So when faced with particularly dry air or particular difficulty in getting water from the soil, a leaf might improve its WUE by reducing the size of its stomatal apertures. Which leaves do. But—and here's the rub—WUE might be better, but inward net diffusion of CO_2 still goes down. And that means a loss of photosynthetic productivity. Adding a minor insult to this greater injury, less outward diffusion of water may mean a higher leaf temperature, which offsets some of the advantage of the reduction in evaporation. Since air holds more water vapor at higher temperatures, internal absolute humidity will be greater, providing more H_2O to diffuse outward. And partial closure may have other undesirable effects as well. The verdict: stomatal control can help—somewhat.

Controlling the Stomata

For the better part of the past two chapters, these stomata ("mouths") have played their part. They not only restrict the connection between the atmosphere and the internal air spaces in leaves but, with their adjustable cross-sectional area, provide the critical control point. As mentioned earlier, even when wide open, they occupy only a tiny fraction of a leaf's surface. A hundred to a thousand per square millimeter sounds like a lot, but their combined cross sections still run around 1 percent of overall leaf surface. The density of stomata is particularly high in the leaves of large trees, while it's particularly low in desert plants. Fully open stomatal apertures are around 6 or so micrometers across and about two or three times that in length. We speak of them as apertures; in reality, though, they're short pipes, around 30 micrometers long.[6]

So how do leaves adjust stomatal openings? First off, the way stomata open and shut has always struck me as mildly counterintuitive. A pair of elongate guard cells, one on each side, flanks each one (grasses, though, do it differently), as in figure 5.2.

Increasing the volume of the guard cells increases the aperture area. If you blow up a cylindrical balloon, it gets fatter; doing this to a pair of guard cells ought to occlude any gap between them. But guard cells have walls of peculiarly uneven thickness—and thus stiffness. Walls right next to the hole are thicker and, as well, have ridges of stiff material next to the top and bottom of the opening. So expansion causes the guard cells to bulge outward and away from the stomatal aperture.

A leaf must continuously adjust the apertures of its stomata as conditions—light intensity plus availability of water and CO_2—change. In the functional world of plants, increasing the volume of cells that have stretchable walls asks nothing especially special. Most of the movements of plants, slow but surprisingly forceful at times, depend on altering the volume of one or another group of cells. Which they do by a process, osmosis, that manages to puzzle most elementary biology students. Garbled accounts notwithstanding, osmosis is really quite straightforward.

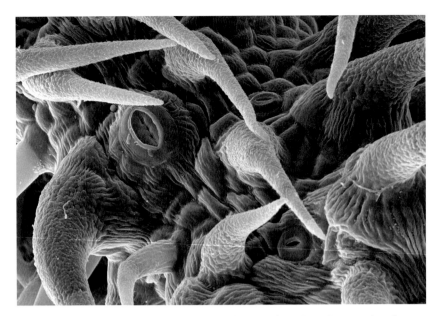

FIGURE 5.2. A photograph of a single stomate (or stoma), made with a scanning electron microscope. To enable the device to get an image, the leaf has been coated with a heavy metal—otherwise the electron beam would go right through it. So, as in all electron micrographs, you're not truly looking at a picture of the object itself, and preparation causes at least a little distortion. Microscopists have labored long and hard to keep images and reality as close as possible. Photograph courtesy of Louisa Howard, Dartmouth Electron Microscope Facility.

Recall that the explanation of diffusion invoked the way molecules move, on average, from places of higher concentration to places of lower concentration. Now consider two compartments of water, one pure and one with a lot of other stuff dissolved in it, as in figure 5.4. Since the dissolved stuff reduces the water's concentration, more water molecules should move from the purer water to the less pure water.

Of course, we're assuming quite a peculiar barrier between the two compartments, a barrier penetrable by water but not (or not very) penetrable by the dissolved material. Peculiar, yes, but ordinary for cells. Speaking of water as being diluted by a solute rather than the other way around may sound odd. But, when you think about it, the notions of dilution and concentration must apply equally to both the solvent (here water) and the solute. Anyway, as a result of the net movement of water into the more dilute compartment, some combination of the volume and

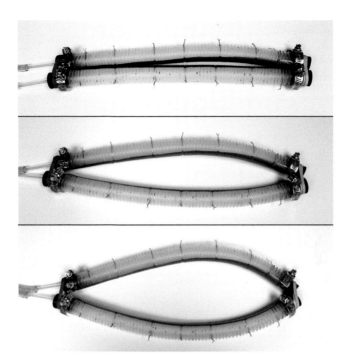

FIGURE 5.3. Pumping up flexible, corrugated hose models the way a pair of guard cells bend outward when the pressure inside them increases, exposing an opening between them. The asymmetry comes from an inextensible lengthwise element on the concave side—here an inch-wide strip of fibrous tape wired to the hose.

Do it yourself: By using the hose of a canister-type vacuum cleaner or hair drier, you may be able to demonstrate, if rather crudely, how that peculiar shape change in the leaf's guard cells opens a stoma. Double the hose back on itself, and apply a length of duct tape to the touching parts of the hose. Tie strings around the hose at the ends of the doubled parts. Secure the duct tape to the hose with additional bands of tape or twist ties. Then connect one end of the hose to the blower output of the vacuum (or to a leaf blower), plug the other end, and turn on the motor. The parallel hoses should bend apart (as in fig. 5.3), exposing a lenticular opening in between.

   In this way, increasing the volume within the guard cells increases the area of the hole through which CO_2 and water vapor diffuse in their separate directions.

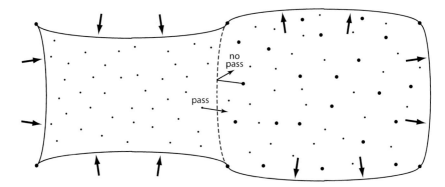

FIGURE 5.4. Here a membrane (*dashed line*) divides a flexible bag into two compartments. The smaller molecules can pass through the membrane; the larger ones cannot. As a result, pressure increases on the side with the larger molecules and decreases on the other side. That makes the bag shrink in volume on one side and swell on the other—a model of what happens when osmotic pressure provides a motor for mechanical change.

the pressure of that compartment—depending on the stretchiness of it walls—must increase.*

Parenthetical note: reversing osmosis means asking water to move

*It's much easier to look at the situation in terms of pressure, assuming no actual expansion, since any volumetric expansion will change the relative concentrations of water and solutes. Osmotic pressure can be either measured or calculated from the composition of a solution. For the latter, one treats the contribution of each kind of dissolved ion or molecule separately and sums the results. For each, the relevant formula is

$$\Delta p = CRT.$$

The formula couldn't be simpler, but the variables have their peculiarities. Assuming SI units, pascals for pressure, C is concentration in what amounts to a count of the individual ions or molecules. That requires correcting for their different masses, done by looking up the molecular weight of each and dividing it by volume; the units are molecular weights per cubic meter. T is temperature in kelvin, not Celsius, degrees. R is the so-called universal gas constant, 8.32 joules per molecular weight (or "mole") per degree kelvin. Sometimes the formula needs adjustment to account for substances that don't completely break up into ions.

Osmotic pressure is one of the so-called colligative properties of solutions, those that depend on the number of dissolved ions or molecules. The other common ones are increase in boiling point, decrease in freezing point, and reduction in a solution's eagerness to evaporate (its vapor pressure). The last two of these, as well as osmotic pressure, have relevance for our leaf.

from a diluted or contaminated compartment to an undiluted, purer compartment. That movement can extract potable (or irrigation-suitable) water from seawater, so a lot of creative energy has gone into the design of equipment for reverse osmosis, and practical machinery is now widely available. Besides requiring an input of energy (about which more below), creating a suitable barrier for a practical desalinization plant presents the main problem. It has to allow passage of water but little else, and at the same time it has to withstand very high pressures. Organisms do such tricks with cells, particularly the outer membranes of cells and their associated gear. But, other things being equal, small compartments can withstand greater pressures than large ones, so cells do something of an end run around the barrier problem.

What needs reemphasis here is the inescapable physical constraint on such a process. Imagine a piston that moved as a result of a volume change caused by osmosis. Water entering its compartment would push it outward. The device would constitute a proper work-producing engine, if a pretty slow one. Here's where basic physics calls the tune. If a device yields work when it does what it prefers to do, then making it do the opposite requires that you put work into it. That's why energy (what it takes to do work) must be supplied to accomplish desalinization. Put in the general terms of the first chapter, desalinization, whether by cells or by utility companies, amounts to unmixing, a nonspontaneous process—like making hot and cold from lukewarm water.

Back to the stomata. To make guard cells swell, the plant must in some way increase the concentration of solute inside—that's how it can persuade water to enter. Cells rarely if ever transport water directly; instead, they adopt an indirect approach, transporting some solute or doing something else that induces the water to do what comes (thermodynamically speaking) naturally. What are the options for guard cells? One possibility consists of breaking starch molecules into their constituent simple sugars, thereby greatly increasing the number of dissolved molecules. That's a well-known mechanism, and it was the first guess as to what was happening. But it's not the whole story. Guard cells appear to transport positively charged potassium ions (K^+ is the usual shorthand) inward. That draws in an equal and opposite charge of negatively charged ions—ones of chloride (Cl^-) and some others. And water just can't avoid following. Every living creature moves water around with this indirect osmotic scheme—I'm sure your kidneys are doing just that at this very moment.

How much pressure do guard cells generate? Enough, obviously, to push outward the epithelial cells surrounding the guard cells on the leaf's surface—themselves pressurized with liquid. Reported pressures range from around 2 atmospheres (0.2 megapascals, in SI units) to over 40 atmospheres (4 megapascals). To put that in context, we pump the tires of our cars up to slightly over 2 atmospheres, and those of our bicycles to several times more. Average blood pressures for humans and mammals smaller than ourselves run around an eighth of an atmosphere. So we're talking about seriously high pressures.* (All the numbers refer to pressures above the background pressure, 1 atmosphere at sea level. We more often speak of pressure differences than absolute pressures, using Δp in formulas even when the idea of difference isn't explicit in our wording.)

I headed this section "Controlling the Stomata" rather than merely "Stomata" as a gentle reminder (to myself as writer as well as to the reader) that the crux of their role consists of control—adjustment of the area connecting air inside and air outside is the means for its task, not the end. The degree to which stomata open depends on quite a number of local factors: availability of water within the plant, illumination level, CO_2 level within the leaf, humidity of the ambient air, local wind speed. "Depends," though, subtly demeans the role of stomata as controllers rather than as mere responders. To make the point, no trivial one, we need to talk about control systems.

First, notice that your kitchen oven has a temperature sensor inside, that your house or apartment has a thermostat on an inside wall, that your toilet (if an ordinary one) has a float on the water in its tank. In each case, the system has a sensory device—for temperature or water level—that provides information. Furthermore, each system features a sensor on the inside. Why should the temperature sensor for a house be on the inside, where little ever changes on its own and where the sensor has the

---

*One rationalization of the high pressures we encounter in small-scale systems goes back to the basic definition of pressure as a force per unit area: $\Delta p = F/A$. (Recall the use of pounds per square inch as a pressure unit, or, better, newtons per square meter, which is what a pascal signifies.) If area decreases and force changes little, then Δp will go up, way up. I'll give a slightly less seat-of-the-pants explanation in chapter 7.

least contact with the outside climate? Why not put it on the outside, poised to tell the furnace when the mercury (anachronistic allusion!) goes down? Simply because the system—and you, the user—really care about the temperature inside, not the temperature outside. If the house gets too cold, for whatever reason except that you're resetting the control, the furnace must be hot to go.

In general terms, by putting the sensor on the inside we're positioning it to be sensitive to the action of the system it controls. It *feeds back* information, a tiny sampling of the system's output, to control its action (fig. 5.5). The system has three critical elements: a motor device or "effec-

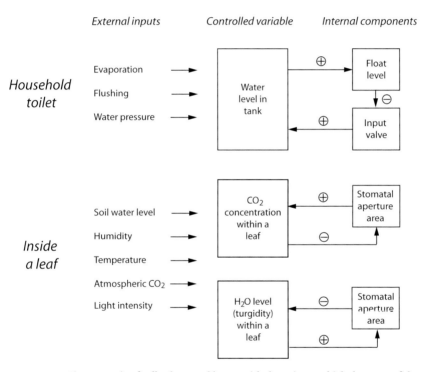

FIGURE 5.5. Three negative-feedback control loops, with the point at which the sense of the information is reversed given by a circled minus sign. The water level in the tank of a household toilet restores itself after flushing and remains constant, even if you leave the top off and water evaporates. A sensor—the float—controls the input valve. If the float drops, the input valve opens until the float gets back to its preset position, and then the valve shuts off. What's critical is that reversal. If the system detects low water, it instructs itself to add water. Leaves control their stomatal apertures analogously, although their system, controlling both water and CO_2, has multiple inputs and interacting set points.

tor," such as a furnace or water inlet valve; a sensor, such as a thermal sensor or float; and an informational link, such as an electrical wire or float rod. Crucially, the system reverses the sense of the information in its decision making. If too cold, add heat; if too hot, heat less (or chill, if the house has an air conditioner). If the water level in the toilet bowl is too low, add water—remove some water from the toilet tank and monitor its refilling. What we're talking about is *negative feedback*, negative not because something is reduced, *but because a signal is reversed*. Every system that holds something constant by taking compensatory action depends on negative feedback. In terms of information, feedback makes the system a closed loop; negative feedback makes it a compensatory system.

(A proper explanation requires mentioning open-loop systems. These are just systems that know nothing of their own outputs, really loop-less rather than in any sense loop-y. They may be elaborately programmed, but they play dumb in the face of changing output conditions. The explanation also requires mention of closed-loop systems with positive feedback. These take action, not to minimize, but to amplify any initial perturbation, such as opening the oven door or flushing the toilet. Where you need compensatory control, they're obviously disastrous: a fire spreads because it heats more fuel to ignition temperature, which heats still more fuel, and so forth. Or a slight reduction in action causes a system to turn itself completely off. Positive-feedback systems do give splendid service as triggering and synchronization devices, of little concern here.)

There's little doubt that, with their guard cells, stomata provide the motors of proper negative feedback, compensatory controls. A lot is known about the immediate (and complex) chemical driver for stomatal opening and closing—the various ions and other entities involved. Somewhat less clear are the details of the various control loops for which the stomata provide motors—the interaction of stimuli, the location and operation of the sensors, and the way the information gets transferred. A few for-instances . . .

Stomata will open in response to a decrease in concentration of CO_2 in the intercellular air spaces in a leaf; that lets in more CO_2, as we'd expect of a well-behaved negative feedback system. The sensor appears to be the CO_2 level in the guard cells themselves, so their own activity forms input as well as motor. Since they're well endowed with chloroplasts, they're proper photosynthetic CO_2 consumers, using the stuff like cells further in. Conversely, at night stomata close (except, of course, in CAM plants), stimulated by the elevated CO_2 of guard cells that now respire but can't photosynthesize any longer.

Complicating the picture, stomata respond directly to light, not just to the drop in CO_2 that comes with increased photosynthetic activity. We know that this activity responds to something other than photosynthesis, because it prefers short-wavelength blue light rather than the broader spectral input that powers photosynthesis.

Temperature provides another stimulus, and it has both direct and indirect effects. At higher temperatures, stomata commonly open further, responding to increased CO_2 consumption and thus lower levels of gas dissolved within the guard cells. But at still higher temperatures a leaf's own respiratory CO_2 production becomes ever more significant, and the stomata begin to close, responding to the higher level of CO_2.

Finally, stomata respond to water, both to the supply of water raised from the ground and to the humidity (and, indirectly, the wind speed) of the air around them. The basic mechanism is just what we expect on mechanical grounds—with either more water around or high humidity, guard cells increase their turgidity and the stomata open. That's also what we expect on functional grounds: with a good supply of water coming up from the ground, low water-use efficiency matters little, and with high humidity, WUE will be high anyway. But here, too, we run into complications, especially when lack of water is the problem. A little water shortage can cause an initial increase in stomatal aperture, most likely because the epithelial cells surrounding the guard cells get softer earlier and provide less pushback. And stomata don't reopen quite as eagerly as expected. Leaves seem to generate a particular hormone (yes, plants as well as animals produce and respond to hormones) when stressed by insufficient water, and persistence of the hormone inhibits stomatal reopening.

Complicating the picture further, stomatal size changes periodically with no external stimuli at all, directed by some internally generated rhythm. Even with no external changes with which they might synchronize, stomata retain a periodicity of roughly 24 hours for a few days. Moreover, some stomatal behavior can't be described as either a simple control loop or an ordinary free-running daily (= circadian) rhythm. For instance, in some plants but not others, stomata open before dawn and close down for a midday photosynthetic siesta—"flower beds," to misuse an expression.

We might compare the leaf's situation with our own, simply by substituting oxygen, our main resource, for CO_2, the chief input for leaves.[7] Air-breathing animals have three advantages over terrestrial plants. The least of these is O_2's slightly greater propensity to diffuse—about 17 percent greater. A basic chemical difference between photosynthesis and respiration matters more. Photosynthesis consumes water, about as many molecules of it as of CO_2. Respiration, in essence the reverse process, produces water at a similar rate: burning 2.1 grams of starch yields 1 gram of water. The biggest difference, though, comes from the vastly different amounts of oxygen and carbon dioxide in our atmosphere. Even with the recent rise in CO_2, there's still over five hundred times as much oxygen around. The difference in aquatic systems is less, because almost thirty times more CO_2 than oxygen dissolves in a given amount of water.* But CO_2 is still less concentrated in water than is O_2 if the water has equilibrated with the air above—500 divided by 30, or a mere 17 times more oxygen.

Animal physiologists don't ordinarily express their data in terms of WUE, but if we want comparisons, conversion is easy enough. If starch is the input and provides exactly enough water to balance supply and demand, WUE is that figure of 2.1 just given. A desert rodent, the kangaroo rat, can do even better, at 2.4, mainly because even dry seeds contain a little water. A lab rat achieves 1.4, a human can manage about 1.6—both leakier creatures that need the occasional drink. Contrast those numbers with a typical value for plants of 0.025—fifty to one hundred times worse. We not only eat plants, but we breathe easy with the fabulously oxygen-rich air they've bequeathed us. To a plant, our water loss problem would be a wet dream.

*One reason we carbonate rather than oxygenate our fizzy drinks; the others, I'd guess, are the legacy of natural fermentation and our taste for solutions with the slight acidity of the carbonic acid generated by dissolved CO_2.

HOUSEPLANTS, EVEN THOUGH they grow only slowly in the low light of our homes, need watering fairly often. Plants demand prodigious quantities of liquid water, often hundreds of times as much as they could possibly use to make sugar from carbon dioxide in photosynthesis. Almost all that water comes from the soil and has to be raised to the leaves. Nothing in physical biology makes a better story than the tale of the ascent of water—mechanically counterintuitive, historically curious, structurally specialized, globally critical. The pumping system has no moving parts, costs the plant no metabolic energy, moves more water than all the circulatory systems of animals combined, does so against far higher resistance, and depends on a mechanism with no close analog in human technology. While a few details remain obscure, we're quite confident that we understand the overall picture—remarkable for something that at first blush sounds pretty bizarre.

First we have to dispose of a mechanism that I see mentioned here and there, mainly in books written by physicists. Capillarity is easy to demonstrate, but it simply doesn't pass quantitative muster. The standard demonstration of capillarity is a fixture of high school courses in physical science, one of those never-fail items on which we teachers rely. One simply inserts a thin glass (i.e. capillary) tube partway into a container of water, and water obligingly rises in the tube, as in figure 6.1.

The narrower the bore of the tube, the higher the water rises. The only potential fly in the ointment is that water has to "wet" the inside of the tube. That is, water must be attracted to the glass surface—which it will be if the glass is clean and thus, as we say, hydrophilic. Grease the glass with some hydrocarbon-based ointment, and the water level in the tube sinks beneath that in the container—the surface has been made hydrophobic. Trees have lots of long, thin hydrophilic tubes running up their trunks, comprising just the right plumbing for capillarity.

How high will water rise in a tube? The water ought to

FIGURE 6.1. Water rises in a glass capillary tube, with the diameter of the tube setting the height it achieves—the smaller the tube, the higher the meniscus. The liquid here is milk, which photographs more satisfactorily than pure water but (from a physical viewpoint) differs little in properties.

rise until the force pulling it up just balances the force pulling it down. Assuming ideal hydrophilicity (a sobering prospect), the force pulling up consists of the surface tension of water (about which a lot more in the next chapter) times the length of the line of contact between the water's edge and the tube. This last, of course, is the circumference of the tube, the line at which the three phases, solid, liquid, and gas, meet. The downward pull comes from nothing more than the weight of water. I'll leave the particulars of the calculation for the next chapter and just declare that at the surface of the earth, the water should rise 7 millimeters in a clean tube whose diameter is 1 millimeter. For a tube a thousand times skinnier, 1 micrometer, water should rise a thousand times higher: 7 meters, or a bit over 20 feet. This last number suggests that we're in the right ballpark for drawing sap upward.

Two problems, though. First, a physical problem. If a tree had conduits thin enough for water to rise to the height of the leaves by simple capillarity, it wouldn't be able to pump much water upward. Yes, water would fill the conduits, but no, it wouldn't flow much at all. Water's resistance to flowing through such tiny conduits would be prohibitively high, making it hard to explain the measured rates at which water leaves the leaves. And then there's the biological reality. The conduits going up through tree trunks don't happen to be that small: their diameters run between about 30 and 300 micrometers (0.03 to 0.3 millimeters). Thirty micrometers sends water only about 1.5 meters (5 feet) upward, and 300 micrometers is ten times worse: 15 centimeters, or 6 inches. Capillarity has a nice physical basis, but it's not what trees use.

Some kind of pressure applied from the bottom—a push—sounds better. Organisms can generate high pressures by increasing the concentration of small molecules in some compartment, like the osmotic pressures described in the previous chapter that run the guard cells. The locally elevated concentration then sucks in water like (but not precisely like) a sponge. Plants, even big trees, clearly have osmotic pumps at the bottom to give water an upward push. Establishing water columns after a winter's freeze and getting the initial, nutrient-rich sap up to developing buds does involve such a push. But for most vegetation most of the time, its role is small.

One drawback to sustained use of an osmotic pump by trees must come down to the supply and fate of the small molecules that draw in the water. Rising sap will move these molecules up a tree all summer (and summer after summer), which requires that they be continuously supplied. Because virtually all the water evaporates at the top, they'd accumulate near the top and reach intolerable levels—if, again, a tree could absorb or synthesize them in sufficient quantities in the first place.

The other problem for any kind of push from the bottom is simple lack of coincidence with observation. A push from below would generate positive pressure. So if a tree were cut into during a summer day, it would bleed fluid like an injured person. In fact, a cut tree, rather than exuding anything, sucks air in with a hissing sound that's detectable with a sensitive microphone. Furthermore, since wood isn't fully rigid—no material is—a tree trunk's girth should respond to its internal pressures. A push from below would make trees swell up a little when water ascended most rapidly. But they do the opposite, shrinking slightly. Shrinking implies a pull, a sucking caused by pressure reduction from above, not pushing with a pressure increase from below.

Capillarity, the obvious mechanism for such a pull, we've already shown to be inadequate. But pull it does, as recognized at least as far back as the Reverend Stephen Hales (1677–1761), among other things perhaps the first quantitatively minded plant physiologist. Hales determined the direction of sap flow, measured both the rate of flow and the rate of vapor emission by leaves, and showed that shoots with their roots cut off could draw water upward against a decent fraction of an atmosphere of pressure.[1]

How High Can Suction Go?

Suction sounds easy enough. We do it all the time with soda straws, and mammalian infants know the trick from the start. Too many insects make a habit of it, doing their nefarious business through our skins and the surfaces of our favorite plants. Yet while suction may look easy, it works only up to a certain height. Ordinarily, neither we nor these insects nor our infants really suck in a strict physical sense. The instances we ascribe to suction inevitably depend on something else altogether. In reality, the sucker reduces the pressure in a compartment (a mouth, for instance), which results in a difference in pressure between the compartment and the outside world. Then outside pressure, now greater, gives a push—what happens is really a push from the excess outside pressure rather than any pull from the inside. When you draw liquid up through a straw, you reduce the pressure in your mouth just a little below atmospheric pressure. So you produce a pressure difference between the container of liquid and your mouth that offsets gravity, and the liquid flows up into your mouth. Strictly speaking, the higher pressure of the atmosphere pushes the liquid upward and does so with a force that depends on the pressure difference.

The limit comes when you reduce the inside pressure to zero—which you can't actually do with either your facial or chest muscles. If you could do so while drinking at sea level, you'd then have made a pressure difference of 1 atmosphere. Working against gravity, that pressure difference of 1 atmosphere could push—from below, remember, even if it looks like a pull from above—water up to a specific maximum height. That height is the pressure, 1 atmosphere, divided by the density of the liquid and by gravitational acceleration. Sucking on water or some palatable equivalent could raise it 10.3 meters, or 33.8 feet, but no more.*

So ordinary pseudosuction, reducing pressure at one end of a pipe below that of the local atmosphere, can't do the job much better than can capillarity. Trees might do an end run around the problem, drawing water upward in a series of stages, lifting it from one reservoir to another. All that the scheme asks is that the hoist between stages be kept below that critical 10 meters. But the requisite reservoirs, devices to repeatedly re-store local atmospheric pressure, can't be found. And they'd be pretty hard to miss. At least one group of trees, the baobabs of Africa and Mada-gascar, do store water in their trunks, water that they then invest in foli-age to get a jump on the next rainy season.[2] So we don't lack a model.

Obviously, trees grow far higher than 10 meters, and the fossil record tells us that they've been doing so for many millions of years. What's the trick? Over a hundred years ago, a botanist and a physicist, Henry H. Dixon and John Joly, made a good case for extraordinary suction, for a

---

*Formula and calculation:

$$h = \frac{\Delta p}{\rho g} = \frac{101,000}{(1000)(9.8)},$$

where Δp is the pressure difference in pascals (newtons per square meter). 101,000 is the pressure exerted by 1 atmosphere, the greatest difference that can be achieved by reducing pressure from what's around us. 1000 kilograms per cubic meter is the den-sity of water, and 9.8 meters per second squared is the acceleration of gravity.

You can build the formula from the basic Newtonian second law, or the defini-tion of force: $F = ma$, force equals mass times acceleration. Consider an erect cylin-der pressing on the ground, and divide both sides of the second-law equation by its cross-sectional area, so $\frac{F}{A} = \frac{ma}{A}$. Replace m by the density of the cylinder times cross-sectional area times height, so $\frac{F}{A} = \rho Aha/A$. The relevant acceleration, of course, is that of gravity, g. Pressure difference is force per unit cross-sectional area $(\frac{F}{A})$, so the original $F = ma$ becomes $\Delta p = \rho gh$, an easy-to-remember rearrangement of the original formula.

Do you want to convert our atmosphere of 101,000 Pa into units more familiar than what SI specifies? Plug those little pascals into the formula, plus the density of mer-cury (13,560 kg/m^2), plus gravitational acceleration (9.8 m/s^2), and you express an at-mosphere as a height in meters of mercury, 0.76. Converted, this becomes the familiar 29.9 inches (760 millimeters), the number that drops when the weather bureau reports an approaching storm.

real pull rather than a disguised push.[3] Dixon and Joly invoked pressures below zero. As they put it, aware of the radical nature of the suggestion, "[this is] the theory which we venture with great diffidence, as to the nature of the phenomena of the ascent of sap." To repeat, they invoked pressures—not just below the ambient 1 atmosphere at sea level but truly subzero. That suggestion meant invoking tension rather than pressure as the two are usually distinguished. However one names the impetus, the system has to pull water rather than persuading the atmosphere push it. Speaking of names, Dixon and Joly's idea has come to be known as the cohesion-tension theory, in case you want to search for more information.

Pull from above rather than push from below explains why a cut aspirates air inward. And it explains the reduction in a tree's girth when sap rises most rapidly. Neither of these phenomena was known to these nineteenth-century pioneers, who took what must be regarded as a brave shot in the dark, an educated guess that what happens is something unlikely but at least minimally plausible. What makes their proposal plausible is that the liquid going up a tree is a liquid.

At this point we have to hit Pause in both our biological and historical accounts in order to say a bit about the behavior of matter in general and of water in particular.

Liquids Are Not Gases

The explanations in chapters 4 and 5 depended on the physical behavior of both gases and liquids. Explanations in this chapter and the next depend on what liquids do that gases don't. For lots of issues the differences between the two are quantitative rather than qualitative, the reason the subject of fluid dynamics addresses both in almost exactly the same words. Both gases and liquids flow, substances diffuse through both of them and squeeze a body of either of them, and both exert outward pressure on the walls of their containers. So how do liquids differ qualitatively, not just quantitatively, from gases? Perhaps the question's focus should be sharpened by asking what keeps liquids liquid, what keeps liquids from vaporizing into thin air. The answer, nothing complex or obscure, holds vast significance for our leaf.

Quite simply, liquids have internal cohesion. Their molecules attract one another, by contrast with gas molecules, each of which goes around (except for the occasional collision) with magnificent disdain for all others. The degree of mutual attraction (plus some other factors) sets the density of a liquid, which then varies only a tiny bit when the overall pressure changes. For a gas, by contrast, density and pressure vary almost indistinguishably—double one and you double the other. In their near-constant densities, liquids resemble solids. You might consider a liquid as either a pressure-indifferent gas or a flowing solid—if that helps.

A solid rod opposes your pull with the mutual attraction (or else en-tanglements) of its molecules. If that's the case, even if you can't pull on a gas, you ought to be able to pull on a liquid rod, working against its solid-like internal cohesion. Therefore liquids should be capable of withstand-ing tension or, as more often put, negative pressure.* That's the physical basis of the prescient suggestion of Dixon and Joly about sap ascent.

Water, H_2O, may be the most ordinary liquid in our immediate world, but when it comes to internal cohesion its behavior is anything but or-dinary. At room temperature and ordinary pressure, chemically similar compounds exist as gases—for instance hydrogen sulfide (H_2S), with sul-fur instead of oxygen; ammonia (H_3N), in which nitrogen replaces the oxygen; and methane (H_4C), with its carbon. Liquid hydrogen sulfide boils at –61°C, ammonia at –33°C, and methane at –164°C, all far, far colder than water's boiling point of +100°C. Despite having molecules over twice as heavy as those of water, ethanol, more than familiar to many, boils at only 78°C. In short, water has a very high internal cohesion.†

Equally unarguable practical difficulties oppose the unarguable logic

*Tension and pressure share the same units, newtons per square meter in SI, tempting us to regard tension as synonymous with negative pressure. That's almost but not quite the case. When you pull on a solid and thus load it in tension, the solid mainly feels (I'm brushing aside a little fishiness here) a unidirectional force. When you pull on a liquid, it pulls equally inward on all the walls of its container. In effect, what you're applying feels likes an omnidirectional force, and it gives an omnidirec-tional response, just as when you squeeze a container of liquid and impel the liquid to leak out of any opening it can find.

†To put a number on it, the internal cohesion of water is around 15,000 atmo-spheres, or 1,500 megapascals. The figure can be derived from water's surface tension, heat of vaporization, or several other properties. Working against this theoretical limit, we could (in theory!) suck water up to a height of 15,000 atmospheres times 10.3 meters per atmosphere, or 154 kilometers—nearly 100 miles. That would be one tall tree!

of Dixon and Joly. Consider a simple physical model, a U-tube longer than 10.3 meters. Force water into it until one arm is full but the other is nearly empty, close the end of the full arm, and tilt the whole rig vertically, perhaps in a stairwell or off the side of a building. What happens? The level in the closed, full arm falls—as does the mercury in figure 6.2. A space, a near vacuum, appears in the top of this arm as the water column refuses to stay up beyond the 10.3 meters of atmospheric push.

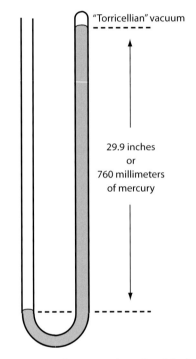

FIGURE 6.2. If one arm, here the right-hand one, of a rigid U-tube is filled with mercury, the mercury will drop until the difference in heights between the two arms is 29.9 inches (760 millimeters)—on a normal day at sea level. We speak of this as 1 atmosphere of pressure and often refer to pressures in terms of the height of just this kind of column of mercury. If it were filled with water, 13.6 times less dense, a column 33.8 feet or 10.3 meters high would be needed—awkwardly large for ordinary use. Note that the volume of the vacuum above the column is irrelevant.

The internal cohesion of liquids may make negative pressures possible, but it doesn't make them ordinary, for at least three reasons.

1. Reducing pressure, what you're doing when you erect the tube, will cause dissolved gas to boil out, as happens when you open a warmed bottle of any carbonated drink. So the water won't stay up even 10.3 meters—the space at the top will fill with water vapor, the amount depending strongly on the temperature (recall chapter 5).

2. Unlike a solid, a liquid under negative pressure must be contained in something to keep it from flowing—again, a liquid, unlike a solid, has no preferred shape. The properties of the container can be troublesome as well—liquid may fail to adhere to its inner wall. The requirement that it adhere to the container is as critical as the requirement that it itself cohere.

3. Moreover, if a column under negative pressure ruptures, how could continuity ever be reestablished? We might suck out air or water, but we can't suck out a vacuum—by definition there's nothing there to suck.

So Dixon and Joly's notion that leaves are somehow sucking sap up the trunks of their trees implies something that is on the face of it unlikely, unstable, and quite wondrous, even if not physically impossible. Mirabile dictu, as the expression goes.

Nevertheless, making a nonliving model turns out to be far from impossible. Our simple straw or pipe may fail, but with sufficient ingenuity, water can be put into negative pressure under controlled, reproducible, definitive conditions. A variety of techniques do the trick. The most obvious (retrospectively!) consists of spinning an open-ended water filled tube (figure 6.3). By Newton's first law, water, having mass, would rather move in a straight line than in a circle. This preference tries to fling it out the ends of the tube, and that stretches the water in the middle, in other words, subjects it to tension or negative pressure.

Lawrence Briggs, of the US National Bureau of Standards, thought up and then tried the scheme back in 1950.[4] From the tube's length, the density of water, and the speed of spinning when the column of water breaks and liquid does fling outward, an unambiguous calculation gives the maximum negative pressure achieved. With this clever device, Briggs

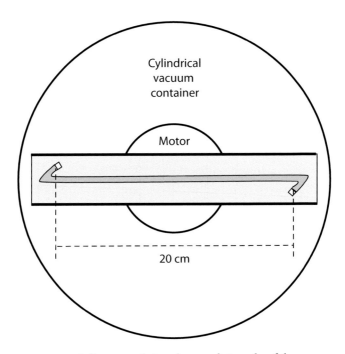

FIGURE 6.3. A diagrammatic top view, nearly to scale, of the apparatus used by Lawrence Briggs to measure the tensile strength of water. The "Z" shape of the tube stabilizes the system—otherwise the slightest offset in the water column would immediately fling water out one end or the other.

produced negative pressures as extreme (maybe one shouldn't say "high") as minus 280 atmospheres. That's enough to hold up a water column 280 times 10.3, or 2884 meters (9462 feet, nearly 2 miles).*

*The calculation is quite straightforward. The pressure tending to break the water column in the center, in pascals, is given by

$$\Delta p = -2\pi^2 \rho n^2 r^2,$$

where ρ is the density of water, about 1000 kilograms per cubic meter; n is the revolution rate, in inverse seconds; and r is the radius of the spinning tube, half the 0.2-meter diameter in the figure. You can easily calculate that the extreme value of 280 atmospheres, 28 megapascals, requires spinning the apparatus at about 22,600 revolutions per minute—quite rapidly, but within the range of fairly ordinary electric motors. Briggs enclosed the tube in a near-vacuum chamber to minimize heating of the spinner by friction with the air as it turns. A closely related formula underlies the design of equipment for dynamically balancing automobile tires.

We can view that negative pressure another way, comparing it to the tensile strength of solid materials. Imagine a rope of liquid water, a square centimeter in cross section, water that can withstand 280 atmospheres, or 28 million pascals, of pull. If it were a solid and dangling downward, one could hang from it a mass of nearly 300 kilograms, a weight of well over 600 pounds. Sure, we have better ropes—steel has about ten times this tensile strength—but this is *liquid water* we're talking about.

Of course, we're also talking about the present record pressure. Still, it's an experimentally determined value rather than some ideal computed from intermolecular forces. Anyway, the datum didn't come easily: nearly all dissolved substances and all possible nucleating solids were carefully purged, the glass tube was scrupulously cleaned and polished inside, and so forth. Lots of times and for no obvious reason, the water column broke at a less extreme pressure. In the end, though, the experiment leaves no doubt that pure water, properly contained, can withstand enormous pulls.

The scheme with which Briggs imposed negative pressures on water isn't the only one that works, nor was his the earliest attempt. Technologists began worrying about the tensile strength—or cohesion—of liquid water almost as early as did botanists. Cohesion plays a central role in the behavior of explosion-generated underwater detonation waves, discovered by P.-E. M. Berthelot (1827–1907), a French chemist, in 1850. And by the end of that century, the tips of ships' propellers reached speeds that occasionally put water into enough tension to break it. Cavitation it was (and is) called, and it makes mischief ranging from lowered effectiveness of propellers to vibrations of drive shafts and entire ships.

Berthelot filled small glass tubes almost completely with water and then sealed them. Heating caused the water to expand more than it caused the outer tube to expand. The increasing pressure inside then forced any residual air into solution, so liquid entirely filled the tubes. Cooling didn't quite reverse the process. Instead, only liquid remained as the tubes were cooled well below the temperature at which the water had originally absorbed all the air. That meant the liquid had to be in tension, which was unmistakably evident when, with an audible click, vapor appeared and the tube's diameter increased slightly. The tubes briefly reached negative pressures of around 50 atmospheres, 5 megapascals. But where did failure happen? In Berthelot's tubes, it more likely came from failure of adhesion of water to glass than from cohesive failure within the water. True cohesive failure, we now know, would have required still more negative pressure. Significantly, he did produce truly subzero pressures, certainly proof of concept.[5]

By the end of the nineteenth century, permutations of the Berthelot technique had minimized uncertainty as to whether cohesion or adhesion had failed. But then and even now the point at which water under tension breaks varies greatly from one experimental run to another, not just among the techniques that produce negative pressures. That points to either randomly occurring nucleating events or variables that have not been brought under control. Recent work may be less relevant to leaves than the experiments of Berthelot, Briggs, and other early investigators. Investigators of explosives interested in the nature of cavitation and in detonation waves have repeatedly measured cohesion during very suddenly applied tensile loading.[6]

Despite a century and a half of recognition, I wonder about an educational system that fails to talk about cohesion in liquids. I can't think how many times I've met an initial skepticism, even denial, as a response to my attempts to explain the accepted view of how water ascends in trees, in particular whether negative pressures can be real, much less generated by trees. And I'm referring not to students but to established physical scientists. (By contrast, biologists assume I know what I'm talking about even when I don't; having one's name on the covers of books has that effect.)

A Less Hypothetical Look

Both the experimental iffiness and the stringent requirements for ultrapure water and perfectly smooth water-attracting pipes raise the critical question: can this supersuction really happen in a tree, and not just occasionally and briefly but routinely and steadily? First, sap composition would have to be fastidiously controlled. But the liquid drawn from the soil does pass through living tissue before entering the main upward conduits. Second, the conduits themselves, xylem, technically, would have to be highly hydrophilic. That does seem to be the case, judging from experiments in which sap-filled twig segments instead of glass tubes are spun. Finally, the conduits, or at least their linings, couldn't be metabolically active, because that would generate dissolved carbon dioxide, which would immediately bubble out and wreck the whole show. But xylem at functional maturity contains no living cells, no more than do our hair and fingernails. All these, though, remain rationalizations, not demonstrations.

At this point what we need are data. What pressures do in fact occur in the xylem conduits when leaves are actively transpiring water vapor? Measuring large pressures doesn't sound all that hard, at least large pressure differences, which are what matter here. An instrument in my lab consisting of little more than a couple of Thermos bottles and some bits of glass and rubber tubing will measure pressure differences down to around a millionth of an atmosphere. Here we're talking about differences of several atmospheres, far more extreme than pressures in human bodies anywhere except in erect penises, and far more extreme than what we measure with very simple gauges in automobile or bicycle tires. So just cut off the end of a small branch and replace it with a manometer to measure the pressure.

But try it, and you run into a disabling practical problem. Cutting off the end relieves the pressure within the branch and leaves nothing worth measuring. Okay—so we add a pump as well as a manometer, and make sure the manometer is connected to the branch with transparent tubing. Turn on the pump, watch the cut end, and record the pressure when the sap first gets sucked out that end. No—that won't work either, since the air in the system simply expands (or a vacuum appears) and the pressure never goes truly negative. You begin to feel the pain of the frustrated experimentalist. The basic problem is that you're asking the pump to pull a seriously negative pressure. None of our pumps can do that under anything like these circumstances. That's where things stood for over half a century, from the original Dixon-Joly paper of 1895 until the 1960s.

Then to the rescue came Per ("Pete") Scholander (1905–1980), deservedly famous among both animal and plant physiologists as the most splendidly ingenious of experimentalists. Scholander designed what has become the standard apparatus, known as the Scholander bomb. Incidentally, the word *bomb* in the lingo of the experimentalist carries no bellicose overtones. It simply refers to a container that can withstand high internal pressure, one in which, by contrast with military bombs, explosion equals failure. For instance, the metabolic energy yield, or calorific content, of a bit of food can be measured by putting a sample in a bomb calorimeter, pressurizing the calorimeter with pure oxygen, and igniting the food by passing current though a resistance wire inside. The food's energy of oxidation appears as heat released as the sample instantly burns—nothing difficult to measure. The pressure in the bomb assures a high oxygen concentration, and its constant volume means that nothing but heat comes out.

The concept of the Scholander bomb is easy enough to grasp and quite obvious with a figure (6.4).[7] To use it, you cut off a small branch that has

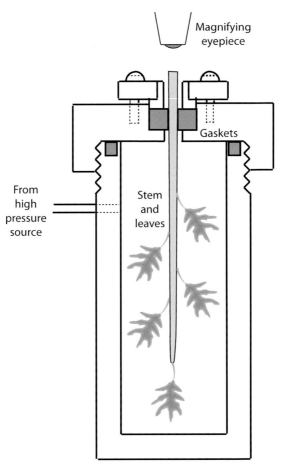

FIGURE 6.4. A longitudinal section through a cylindrical Scholander bomb redrawn from the illustration in the original paper by Scholander et al. (1965). This has to be quite a sturdy piece of equipment, given the pressures involved, so in operation it won't behave as a military bomb would. These days you can't call it a bomb if you mean to ship it by common carrier!

some leaves and immediately put the leafy end, not the basal end, into the bomb. The cut end sticks up out of the bomb through a good airtight seal. Relief of the negative pressure by your cut has allowed the branch to expand, imperceptibly to the eye, but enough to draw the sap down into it. You then apply lots of positive pressure to the bomb from some tank or pump—nothing limits how severely positive a pressure can get. While you put this squeeze on the outside of the branch you watch the cut end with a low-power microscope. When the applied pressure pushes the sap back to the cut end, record the pressure. Now the datum isn't exactly the negative pressure we asked about. Strictly speaking, it's the pressure needed to restore the status-quo ante. The pressure is high outside (the branch, not the room) and lower inside, as ought to have been the case before the branch was amputated. Put another way, the Scholander bomb applies the positive pressure necessary to counterbalance that original negative pressure. So the positive pressure in the bomb should be precisely the same in magnitude and opposite in sign as the original negative pressure in the sap.

(This kind of null-balancing approach to measurement has a long and glorious history in experimental science. Instead of trying to measure something directly, one contrives a scheme to offset that something, and then one measures whatever was necessary for the counterbalancing manipulation. It lends itself to simplicity and accuracy, at least where the phenomenon in question persists long enough for balance to be achieved. It's exactly what we do when we weigh someone by sliding the compensating weight of a scale until its beam balances.)

Particularly interesting pressures should occur in the twigs and leaves near the tops of tall trees. Scholander, famous for field-friendly schemes, didn't take hundred-foot ladders into jungles. At least initially, he employed sharpshooters to cause twigs to part from branches; fallen fresh twigs were then popped into his bomb.

Scholander bombs have now been in use for about half a century, and loads of data have accumulated, obtained from all kinds of plants under all kinds of conditions. What pressures do they report for the xylem conduits of trees? Counterbalancing the negative pressures in the conduits usually takes extremely high pressures, well beyond what offsetting gravity requires. By the argument so far, pulling sap up a 30-meter tree should take about 3 atmospheres of negative pressure—one for each 10.3 meters of height, less any positive pressure contributed by push from the roots. But bomb data come out much higher: 10 atmospheres is nothing; 20 is perhaps typical; the record to date, I think, is *120 atmospheres.* Human blood pressure runs 10 to 20 percent of an atmosphere, so these pressures approach a thousand times what our hearts can produce.

Now experimentalists come to expect asymmetry—errors on the low side most likely reflect apparatus troubles such as leakage, but errors on the high side can't be so easily explained away. (More often than not, bad calibration is to blame.) So we're confident that the bombs measure something that's really there, despite the extreme values they produce and despite (we must remind ourselves) their indirect approach of measuring not negative pressure but the positive pressure necessary to offset it.

Nonetheless, some questions remain. Is the pressure in the bomb offsetting just the negative pressure of the xylem, or is that positive pressure being invested somewhere else in a stem as well, with only a fraction of it trying to push sap back toward the cut end? On the one hand, as just noted, low-side errors bother us more than high-side ones, and this would be a high-side error. On the other hand, from long experience we suspect fabulously extreme numbers of being quite literally fabulous. Confidence was shaken for a while by data obtained with a novel kind of pressure probe. But it was restored in the mid-'90s by two almost simultaneous reports of work by reputable people in the best of journals.[8] Both groups did the Briggs trick, not on glass tubes, but on freshly cut branches of real trees, spinning them until their water columns broke. One group used a variety of woods, recording pressures as low as –35 atmospheres (–3.5 MPa) for the tension it took to break a sap column in a juniper. The other group found pressures down to –19 atmospheres (–1.9 MPa) in western redbuds as well as near-perfect coincidence with the compensating pressures measured on adjacent branches with a Scholander bomb. Sighs of relief all around.

You might well ask what these extremely negative pressures do to evaporation within leaves. No matter how negative the pressure, water evaporates from a free surface just as avidly. Just when you feared yet further complexity, a blessed element of simplicity creeps in, like a snatch of melody interrupting a persistent cacophony.

But we're not home free quite yet. What happens when the sap in a tree freezes? Gases are less soluble in ice than in cold water. We're all familiar with the way water outgases when it freezes, even if we care little about it. Just recall how in an ordinary refrigerator ice cubes come out filled with bubbles. A bubble of gas in xylem conduit represents an instant disaster—with the liquid under negative pressure, the bubble will immediately expand until a near vacuum fills the entire conduit. So it would be a rare conduit that doesn't embolize during the winter. But somehow, come spring, all the conduits fill again.

Other such embolisms appear more or less randomly during the growing season; by fall, the majority of conduits in most trees may have embolized. Since leaves keep on emitting water, somehow ascending sap can circumvent embolized conduits. That means an embolism can in some way be kept from spreading from root to leaf or from conduit to parallel conduit. Finally, why are the negative pressures so extreme? They're almost always several times greater than what would be needed to hold up a column of water the height of a tree. For some such issues, we have good answers; for others, considerable uncertainty remains. Fortunately for the job security of practicing scientists, science always remains a work in progress.

Bottom line: evaporation from the walls of leaf cells into the air spaces within leaves draws water up from the tree roots through conduits that form a sheath around trunk and branches, a sheath located at the interface between the outer bark and the inner wood. Stephen Hales certainly understood that. And the great nineteenth-century plant physiologist, Julius von Sachs (who, unsurprisingly, misunderstood the sap ascent mechanism), encapsulated the overall process as well as anyone since. Writing in 1882, he said, "Thus, by means of evaporation or transpiration, a continuous flow into the organs of assimilation is rendered possible."[9]

Moving Sap

We started by wondering how trees could hold up a column of water or sap higher than what gravity and atmospheric pressure ordinarily permitted. The explanation looks good, but the measurements show, paradoxically, that trees do far better than what the explanation requires. The tallest tree, around a hundred meters high, should need only 10 atmospheres of negative pressure, yet values of 20 and more occur routinely. Resolving the paradox requires yet another element of real-world physics. A tree doesn't just hold up a continuous column of sap from root to leaf, it moves the sap upward, supplying a little water for photosynthesis and a lot for evaporation. It moves sap upward through relatively narrow pipes, pipes with diameters between about 0.03 and 0.3 millimeters, a thousandth to a hundredth of an inch.

Recall the no-slip business from chapter 4. At the walls of such a tiny pipe, as with a pipe of any other size, the speed of flow will be zero. The speed rises, reaching a maximum in the middle of the pipe, as shown in

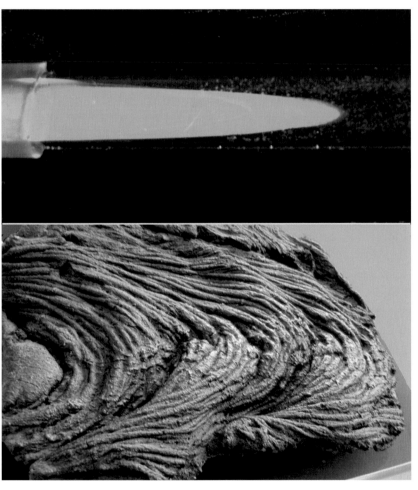

FIGURE 6.5. Two instances of parabolic flow. At the top, a slug of fluorescein dye moves slowly through a glass tube that's half an inch in internal diameter; at the bottom, preserved when it cooled and hardened, lava (pahoehoe form) oozed through some kind of channel (photographed at the museum at Volcanoes National Park, Hawaii).

figure 6.5. So a gradient of flow speeds will extend across each one of these pipes, and the gradient will occur everywhere along its entire length. Like all fluids, sap has viscosity, which means it doesn't take kindly to cross-flow speed gradients. Thus energy is needed to maintain these inevitable gradients. Newton's first law may say that once moving, an object continues in motion with no further urging, but that law presumes no friction—or its fluid equivalent, viscosity.

As a result, sap in a conduit needs a continuous push or pull to keep it moving. The only kind of push or pull that can be transmitted along a pipe's length is pressure. As a result of an applied pressure, a fluid will flow from a place of higher pressure to one of lower pressure. In other words, pressure must drop in the direction of flow. So the energy source that makes sap move appears as a lengthwise pressure gradient.

The pressure gradient needed to hold sap up, again, is 1 atmosphere, or 101,000 pascals per 10.3 meters—roughly 10,000 pascals per meter. What's the additional pressure gradient needed to keep the sap moving? We can calculate it from the diameter of a typical conduit, the speed of movement, and the viscosity of the liquid. The relevant rule is the Hagen-Poiseuille equation (named for the near-simultaneous originators), which often is known as simply the Poiseuille equation.* Incidentally, Jean-Louis-Marie Poiseuille (1799–1869), like Fick from chapter 3, was a physiologist—our efforts to understand how life works have repeatedly stimulated recognition of novelties in physics, just as efforts in physics have required novel mathematics. The equation works for fluid movements

---

*The Hagen Poiseuille equation is

$$Q = \frac{\Delta p \pi r^4}{8 l \mu},$$

where Q is the total flow (volume per unit time, or cubic meters per second in SI), r is pipe radius, l is the length of pipe under consideration, and μ is the fluid's viscosity. Nothing tricky, but notice that for a given pressure difference, Δp, driving it, total flow varies with the fourth power of the radius (or diameter) of a pipe. Halve the bore, and flow will drop sixteenfold! (For turbulent flow, a different and slightly messier formula has to be used.) We're interested in pressure drop per unit length, $\Delta p/l$, so the equation can be made more useful by converting total flow, Q, to flow speed, v. Since $Q = v$ times cross-sectional area, πr^2, then, after rearranging,

$$\frac{\Delta p}{l} = \frac{8 v \mu}{r^2}.$$

slow enough that go through pipes small enough so that flow remains laminar—conditions easily met here.

The distinction between laminar and turbulent flow that came up in chapter 4 needs a few more words at this point. All but the gentlest of winds blowing across a leaf will be turbulent (however it appears to the leaf), but all flow through the internal pipes of plants will be laminar. That sounds—as seems intuitively reasonable—as though air (or gas) moves turbulently and water (or liquid) moves laminarly. But either kind of fluid can move in either mode. We have a reliable (if not absolutely reliable) rule that tells us which mode of flow should happen where. It's another of those dimensionless indices that compare two kinds of forces, an analog of the Péclet number discussed in chapter 4. This more famous one compares inertial force, what keeps a bit of fluid flowing once started, with viscous force, the internal friction of velocity gradients that slows down any flow near a surface. Put another way, it compares the individuality with the groupiness of bits of fluid. As mentioned earlier, if bits behave like individuals in a milling crowd, the flow is turbulent; if they march in lockstep, the flow will be laminar.

The index goes by the name Reynolds number, after Osborne Reynolds (1842–1912), first professor of engineering at Manchester University, who did the now-classic experimental work that established it as a general rule. The index contains the product of speed and size. For size, we take some convenient (or conventional) length, ignoring the details of shape— this isn't a high-precision rule. So big plus fast means a high Reynolds number and turbulent flows, while small plus slow means a low Reynolds number and laminar flows. In biology, small almost inevitably implies slow, big goes with fast, and the range of relevant Reynolds numbers is huge—perhaps a (US) trillionfold between a tiny diatom sinking in the ocean and the whale that swims past it. The Hagen-Poiseuille equation just mentioned works only for laminar flows.* It's great for sap and blood,

---

*The Reynolds number (Re) contains two variables besides speed and size; unsurprisingly, these are density (related to inertia) and viscosity (for viscous force). Nothing complicated, though, as

$$\text{Re} = \frac{\rho v l}{\mu}.$$

The Hagen-Poiseuille equation, with the pipe diameter as the characteristic size, is good up to a Reynolds number of about 2000, that is, to the point where flow in a cir-

inapplicable for household plumbing. We'll meet the Reynolds number again in later chapters.

Back to sap flow. It's easy to measure the size of xylem conduits from low-magnification cross sections, and the viscosity of sap differs little from that of pure water. Determining flow speed takes a little more ingenuity. Sachs, back in the 1880s, supplied a weak solution of lithium nitrate in place of normal sap. Lithium changes the color of the flame (flame photometry) when a stem containing it is burned, so he could see how far the solution had traveled in a given time. These days radioisotopes do that kind of job without the fire, although governments tend to look askance at their casual or outdoor use. A common way to measure how fast sap ascends consists of applying a pulse of heat through a resistance wire wrapped around a tree trunk. A sensitive temperature detector mounted higher up the trunk picks up the pulse a little while later; the elapsed time and distance between heater and sensor tell how fast the sap is moving.

Okay, how fast *does* sap move? Not surprisingly, data vary enormously, depending on type of tree, time of day and year, atmospheric conditions, and other factors. Flows for a deciduous tree such as an oak, with vessels up to about 0.3 millimeters in diameter, run up to roughly 40 meters per hour, or about 10 millimeters per second.* That's an inch every two and a half seconds, more than ten times the speed with which blood is flowing through your capillaries as you quietly read this.

Flows for conifers such as pines never reach the speeds recorded for many of the deciduous trees; a fast flow for a pine tree with conduits around 0.05 millimeters in diameter might be around 2 meters per hour, or 0.5 millimeters per second. Some vines have especially large pipes, not uncommonly around 0.5 millimeters in diameter, and their fastest flows are especially fast. One smells a rule—the larger the conduit, the faster the maximum sap flow in it—nor does the odor mislead, even if it's only the roughest of guides.

How much pressure does it take to drive such flows? Consider a typical midsummer-day speed of 10 millimeters per second through a conduit

---

cular pipe becomes turbulent. That's for pipe flow, not for all situations, such as flow across airplane wings and golf balls.

*That gives a Reynolds number of 3.0, well down in the laminar range for which the Hagen-Poiseuille equation works. For the conifer in the next paragraph, the Reynolds number is a mere 0.025.

0.3 millimeters across. Hagen and Poiseuille tell us that the pressure loss is about 2000 pascals, or about 2 percent of an atmosphere of pressure per meter of tree height. In context, that datum carries significance. We figured earlier that pressure should drop on account of gravity by 1 atmosphere for every 10.3-meter increase in height. That's 9800 pascals of pressure per meter of height. Bearing in mind the roughness of the estimates, the 2000 pascals per meter needed to keep sap flowing is lower but of the same order as the 9800 pascals per meter extracted by the earth's gravity. For the same flow speed in a smaller vessel, say 0.2 millimeters across, the pressure loss would be greater, about 4500 pascals per meter. For the pine mentioned earlier, it would be 1600 pascals per meter—again in the same range.

So what? We might look at the situation from the point of view of a designer, whether a human doing a cost-benefit anticipatory analysis or natural selection doing what amounts to the same thing through its perpetual post-hoc pruning. Fewer, wider conduits would reduce the cost of pulling fluid along. But they would also reduce the safety-in-numbers assurance that water transport upward won't be interrupted by failure in the conduits, failure that's all too likely given their prodigiously negative internal pressures. Any little ding and a xylem conduit may pull a vacuum (cavitate), rupturing the critical continuity of the water column, and thus emptying the whole thing lickety-split. Not only are larger conduits more prone to embolize than small ones, and not only is an embolism in a large one more consequential, but repair of embolisms seems to be easier in narrower conduits than in wider ones.

How wide should the prudent tree make its piping? The wider the conduit, the lower the pressure needed to impel the sap upward. But no conduit can reduce the overall pressure drop below 9800 pascals per meter unless it can mess with the earth's gravitational acceleration or with the density of water. So maybe trees make their conduits wide enough to get the cost as pressure loss from flow down into the same range as the cost as pressure loss from gravity. Conduits wider still would give diminishing returns in return for their risk. What profit would there be from building a structure that's much sturdier than its foundation? Conversely, making conduits smaller means greatly increasing pressure loss from flow, and that means still more negative pressures and thus greater risk of embolism.

At the same time, we can discern another loose rule. Not only do conduit width and flow speed change in tandem, but both change the same way that overall water use changes. Really wide conduits appear in vines and broad-leafed trees, which generally use (or lose, depending on your attitude) water at high rates; narrower conduits occur in the less thirsty conifers.

> **Do it yourself:** You, too, can show how fluid will flow through pipes extending the length of a bit of woody twig. Red oaks provide especially good kinds of twigs for the demonstration; these are the oaks with pointed lobes on their leaves. Woody grapevine is even better. Just cut an inch or two of twig, trimming the ends with a very sharp knife if your pruner crimps the vessels. Blowing through it with the far end in a container of water produces lots of small bubbles. More impressive, if no more revealing: attach a bit of twig to a gas jet, turn on the gas, and light the far end.

It's a good story, but once again, continuing with water, drizzle descends on the parade. Do the losses of pressure from the flow of sap correspond to what the Hagen-Poiseuille equation predicts?

Without notable heroics, we can take a cut piece of trunk or branch and force fluid through it while keeping track of both the forcing pressure and the rate of flow. Sometimes the equation accounts quite nicely for the relationship between the two. But all too often, specimens have considerably more resistance than what it predicts and so absorb more pressure—according to measurements that go back well over half a century. For vines such as those of grapes, the equation works almost perfectly. For some of the oaks, it gives only a slight overestimate of flow (or underestimate of pressure loss, the same thing). For a lot of other hardwoods, such as beeches, birches, and willows, flows are only about a third of what the equation predicts. For shrubs, it's even worse.[10]

It turns out that one more element figures into the pressure needed to drive the flow. The conduits we've been talking about don't extend as unbroken pipes of constant bore from roots to leaves. More often than not they're interrupted at intervals by various sorts of perforated diaphragms, as shown in figure 6.6. At their locations, these arrays of pores greatly re-

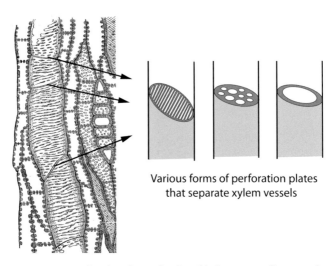

Various forms of perforation plates
that separate xylem vessels

FIGURE 6.6. Perforation plates of various kinds connect adjacent ends of xylem vessels in hardwoods. All impede flow to some extent. The drawing on the left comes from Julius von Sachs's great book on plant physiology, written in 1882.

duce the cross section available for upward flow. The interruption intervals vary widely from one kind of tree to another—in red maple about every 12 millimeters, in holly about every 1.3 meters. (Measuring vessel diameters may be simple, but measuring lengths is anything but. Still, data do exist.)

We'll see in the next chapter how such plates limit the propagation of embolisms and how they play a role in any subsequent refilling. Here our concern is the additional pressure loss they impose. The fluid mechanists have given us decent formulas for pressure loss for flow through circular apertures, although they get less attention than the ones for flow through pipes.* The two equations for loss in pressure due to flow through pipes and through pores, together with the relevant anatomical data, predict pressure losses a lot closer to those caused by gravity.[11] Bottom line: rais-

*The analog of the Hagen-Poiseuille equation, fortunately with no name attached that might challenge native English speakers, looks a lot like it:

$$Q = \frac{r^3 \Delta p}{3\mu},$$

with r now the radius of the pore—essentially a pipe of zero length—instead of the diameter of a pipe. It works up to Re = 3, above which flow becomes turbulent.

ing water to the top of a 30-meter tree takes more than the 3 atmospheres that gravity demands—by our accounting so far, about twice as much. As we'll see in the next chapter, yet another factor takes its cut of the pressure pie as well.

Whatever the specific numbers, we shouldn't lose sight of just how strange a picture we're painting. Every tree you see that's more than perhaps 10 feet tall contains a host of pipes, each of which on a nice sunny day does something unknown in practical human technology, something on the face of it quite implausible, something tickling the unstable edge of destructive cavitation. The energy to run the show comes not from any metabolic chemistry but directly from the environment: it's a solar engine based on the vaporization of water in the leaves, and it needs no moving parts.

At any given time, many of the pipes don't work—cavitations have occurred, leaving them with emboli rather than columns of sap. At least that's the conclusion when we do the numbers, cranking in the measurable rate at which water leaves the leaves, the measurable speed of flow upward, and the total cross-sectional area of the vessels of the trunk. The experimentalist—this author, for one—finds great comfort in the tidy correspondence between the physical arguments, the calculations, and the reality of measurements, however strange the system. Maybe it's not as weird as quantum mechanics, but it's about as weird (and wonderful) as physical biology ever gets.

OUR LEAF FORMS the terminus of a system of pipes that ascends from the roots of the tree, a system containing nearly pure water under fabulously negative pressure. With such narrow pipes, the leaf doesn't risk being sucked into the system, so that's nothing to be concerned about. We all know that an unaided finger can cap off a narrow vacuum tube and remain uninjured—despite our soft skin. But the system presents another peculiar problem, one far from trivial, one that initially appears far from amenable to any obvious fix.

Quite simply, it's one thing to hold up a closed column of water against gravity, a column extending upward to a height well beyond what atmospheric pressure willingly supports. But how can the trick be done in a column that's open at the top? If water can evaporate out of a leaf, then in some nontrivial way the columns of water we've been talking about must be open at the top. Never mind the solid material of leaves—what keeps air from being sucked in and filling the system? In an ordinary open pipe, that will happen at pressures far below even atmospheric pressure. And –20 or –30 atmospheres represents an almost unimaginable sucking strength.

To deal with this problem as well as several related issues, we need to worry first about how curved surfaces resist pressure differences, and second about the behavior of any interface between air and water. The first turns on a basic geometric relationship; the second depends on a physical phenomenon, surface tension, which plays several critical roles in the life of a leaf.

About Curved Surfaces

Let's start with something close to home. You may notice (or can easily find out) that the skinny tires of racing bicycles can be inflated to very high pressures—around 100 pounds per square inch, almost 7 atmospheres or, in SI, the better part of a million pascals above the pressure of the outside atmosphere. Your larger automobile tires take only about

35 pounds per square inch, around 2 atmospheres, or a few hundred thousand pascals. The still-larger tires of trucks and construction machinery require still-lower pressures. Obviously, the bicycle tires don't have especially thick walls—quite the contrary. Conversely, the truck tires have impressively sturdy sides and treads. It's all a little odd.

Still odder is a cute and counterintuitive trick you can try for yourself. An interconnected pair of balloons resists simultaneous inflation, as in figure 7.1. Since interconnection ensures that their internal pressures are

FIGURE 7.1. A pair of balloons on a Y-tube. Blowing into both simultaneously inflates only one, as on the left. To get both inflated, as on the right, they have to be blown into one at a time. That's easy enough with the three-valve setup shown, but tricky otherwise. If both the lower valves interconnecting the two inflated balloons are opened, one balloon will usually dump most of its air into the other. The valves and other fittings shown here, satisfactorily inexpensive, normally connect garden hosing.

the same, you would expect both balloons to inflate. Since the pressure inside each must be the same as that inside the other, whichever inflates furthest will resist more strongly and deflect air toward the smaller, limper one. But no—the smaller one is limper and the larger one tauter, yet the smaller one better resists further inflation. It looks as if the size of the balloon influences the way pressure stretches the rubber.

Only a minor part of this strange behavior depends on the mechanical eccentricities of rubber. It mostly goes back to two more basic issues.

Do it yourself: It's easy enough to arrange some sort of T-tube or Y-tube using cutoff valves from hardware store parts (as in fig. 7.1). Check the plumbing department or garden shop for the fittings. Attach spherical balloons to two of the ends. After you're satisfied that the balloons will not inflate together, clamp off, plug, or tie the input tube. Poke each balloon to get a feel for the tension in its wall. Then cup your hands around the larger one and squeeze it. Initially it resists strongly, but then as it gets smaller it gets more accommodating, finally collapsing as it empties most of its air into the other balloon, making the latter now the larger. Squeezing this one should do just the same thing, restoring the way things were before.

First, pressure doesn't stretch a balloon (or inner tube), at least not directly. Tension, as a force in the plane of the balloon's surface, does the stretching. Then there's the peculiar relationship among three variables: the curvature of a surface, the pressure difference across the surface, and the tension in the wall generated by that pressure difference. We might reasonably guess (if for some odd reason the issue came up) that a given internal pressure should produce a given tension in the curved wall of a container. Not so—the relationship between the two involves an odd scaling rule. The rule emerges not from physics but from geometry; it was uncovered early in the nineteenth century by (at least) two polymaths, Pierre-Simon Laplace (1749–1827) in France, and Thomas Young (1773–1829) in Britain. Biologists (those who know about it) call it Laplace's law or sometimes the Young-Laplace equation. Physical scientists, with other things memorializing Monsieur Laplace, don't ordinarily put anyone's name on the rule. (Young will reappear later; the colorful polymath gave us, among other things, Young's modulus, which happens to describe how the material of the balloon responds to the pressure-induced tension.[1])

An analogy might head your intuition in the right direction. Imagine a rope around a bundle of cylinders, as in figure 7.2a. Pulling the rope taut snugs the cylinders against each other. If the bundle contains only a few cylinders, as on the left, then a modest pull gives a lot of snugging force—the rope curves far enough around each one so that its net effect is a strong center-directed pull. If the bundle contains a lot of cylinders, the

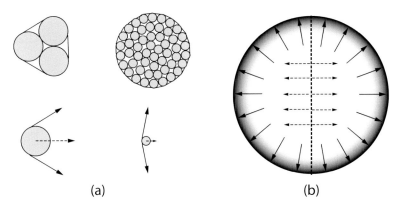

FIGURE 7.2. Two ways to introduce Laplace's law. In (*a*) a given tensile force in a rope surrounding a few cylinders produces a greater squeezing force on the bundle than does the same tensile force in a rope around a larger number of cylinders. In (*b*) an outward pressure on the wall of a sphere or cylinder generates tension along any line going around the sphere or cylinder.

same pull gives less snugging, as on the right. Now imagine that all cylinders, left and right, are the same size, so the bundle on the right is much larger. Same pull, much less snugging of the right-hand one, as before. (You might try this with loops of thread and bundles of spaghetti.) Cylinders—they're molecules; pulling on the rope—that's tension; snugging the cylinders—that's pressure. A given tension on the outside of a bigger cylinder gives less pressure within. Put the other way, a given pressure within a bigger cylinder gives more tension on its outside. Spheres behave the same as cylinders. Bigger means less pressure resistant, other things being equal.

Now to move a little closer to reality, imagine dropping the separate cylinders and replacing the bundles with a thin-walled, hollow sphere, one that's kept inflated by a pressure inside at least a little greater than that outside. Tension within its wall holds the sphere together, as in figure 7.2*b*. Thus if you were to cut it in two around its midsection, relieving the tension, the two hemispheres would be blown apart by the pressure. Tension obviously resists the pressure, and a cut would drop that tension to zero.

What if the hollow sphere were doubled in diameter without altering the pressure inside? The inner pressure force, acting on the internal wall area, will increase—not twice, but four times. That's because when you increase the diameter, radius, or circumference of a sphere by some factor, you increase its surface or cross-sectional area by the square of that factor.

Tension, by contrast, operates along a line—such as the line of the cut around the middle of the sphere. Doubling diameter thus increases tensile force twice, not fourfold. So the bigger the sphere, the more effective pressure becomes in generating tension in its wall. And so the tension in the bigger sphere (before our cut) was twice as great as that in the original one: with no pressure change, doubling the diameter (or radius) doubles the tension. Yes, the larger sphere will feel stiffer than the smaller one, but what you're really feeling is the tension in its wall, not the pressure inside.

In this functional sense, the larger sphere has the weaker wall. To withstand the same pressure, its wall has to be either thicker or stronger. Otherwise, it has to be operated at lower pressures. Similarly, a smaller sphere with the same walls can withstand a higher internal pressure. Bear in mind that we've said nothing about the material of the sphere—the argument is purely geometric.* Material enters the picture only when we worry about how the sphere handles the greater tension. The argument holds for both spheres and cylinders, even for cylinders that happen to curve around and form toruses (tori, if you prefer), like tires. Does the contrast between skinny and fat tires now seem at least a little less odd?

Back to the paired balloons and logical closure. For the sake of argument, assume that this balloon doubles in diameter, radius, and circumference with little change in internal pressure, as in figure 7.3. The tension

*Formal formulas for Laplace's law:
For a sphere, which curves in two directions, pressure gives a particularly effective squeeze,

$$\Delta p = \frac{2T}{r} \text{ or } T = \frac{\Delta pr}{2}.$$

For a cylinder, which curves only one way, pressure does less, so the 2 becomes 1.

$$\Delta p = \frac{T}{r} \text{ or } T = \Delta pr.$$

T is wall tension in the plane of the surface, in newtons per meter of length, with length perpendicular to the pulling force. Bear in mind the slightly unfamiliar units for tension, which nonetheless keep the equations dimensionally correct. An engineer would convert tension to tensile stress, a concept we don't need here, so the formulas would look a little different.

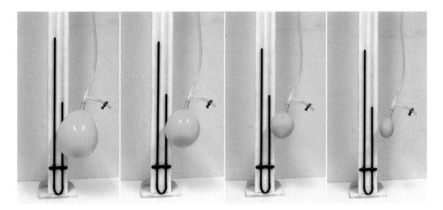

FIGURE 7.3. A balloon in various stages of deflation (inflation would just reverse the sequence) connected to a manometer. Pressure inside varies a lot less than you might expect. But consider: if you can start a balloon inflating, you can blow it up as far as you wish. It doesn't fight back more and more as it enlarges. The manometer fluid, incidentally, is water with some blue colorant.

in the wall will double. Your blowing, then, disproportionately affects whichever balloon is larger. Its inflation keeps the pressure from rising further as it itself gets, in this odd sense, weaker and weaker. Again, applying pressure doesn't directly cause inflation but does so by stretching walls which, in turn, do the job.

[Digression. As mentioned, the same thing happens in cylinders. You're undoubtedly familiar with the odd way a cylindrical balloon inflates, one part expanding almost fully and the expansion then creeping along its length. Only a peculiarity of the response of rubber makes cylindrical balloons even possible. If rubber stretched in the simplest manner, with the stretch increasing exactly in proportion to the force, then whatever part of the balloon inflated first would expand and burst before the rest of the balloon inflated at all. If that happens in a blood vessel, we call it a ruptured aneurysm and declare a major emergency. Normal arterial walls have a peculiarly aneurysm-resisting structure that allows them, unlike balloons, to expand uniformly when your heart beats. By contrast, some inferior quality hoses used to connect clothes washers to household plumbing can develop aneurysms. We've had two such cases in our house, one of which ruptured while we were away and a neighbor was checking the place daily. As her granddaughter, tagging along, put it: "Grandma, it's raining inside."]

About Surface Tension

The previous chapter said disparaging things about surface tension. While the rise of a liquid in a narrow tube (capillarity) depends on it, chapter 6 made the case that capillary rise pales into triviality faced with the problem of hoisting water up a tree. That inadequacy shouldn't be taken as a general dismissal—surface tension remains a force to be reckoned with. Without it a tree would be out of business on several accounts. Surface tension is yet another item that's unfamiliar in our perceptual world but takes on a critical role in our present story. So we need a look at its origin and consequences.

Liquid water, again, has internal coherence—its molecules know and like one another, unlike gas molecules, which when they meet just mindlessly collide and bounce apart. That mutual attraction pulls a molecule far from a surface in all directions at once, so the attraction mainly serves to keeps a liquid in its liquid state. But the situation of a water molecule at a surface differs. Its peers pull it back into the body of the liquid while at the same time molecules on the other side of the surface to which it's exposed pull it more strongly, less strongly, or not at all.

If the exposed surface of water meets a solid to which its molecules adhere less strongly than their internal coherence, then its edge creeps away, and we say that the surface is hydrophobic. By contrast, if on the outside water meets a solid to which its molecules adhere more avidly than they cohere among themselves, the water's edge creeps around it—as with water's capillary rise in a clean glass tube, whose inner surface we then say is hydrophilic. As mentioned in the previous chapter, the cohesion-tension model for the ascent of sap demands that the inner surfaces of conduits be highly hydrophilic.

Wherever the cohesive attraction among water molecules exceeds their adhesion to any foreign material, bodies of water try to minimize their exposed surface area. If on the outside water meets a gas or a gas mixture such as air, the water is left to its own devices, affected only by its own coherence. So at the air-water interface, water molecules experience a fully unbalanced inward pull—every surface molecule prefers to be fully submerged. Viewed on a larger scale, we see that a bit of water rounds up into a droplet in order to minimize its surface area. The surface-minimization force is what we recognize as surface tension. The values of surface tension usually cited apply to this extreme case, that of a purely inward pull, as happens when water meets air.*

> **Do it yourself:** You can observe this minimization of surface area by separating two water-air interfaces that have different values of surface tension and watching how they interact. Place a loop of sewing thread, perhaps 10 centimeters (4 inches) in diameter, on the surface of clean water in a large bowl. The thread will almost certainly not form a regular circle. Then touch any part of the water's surface within the loop with a bit of soap or detergent, and watch the thread become circular. The soap has reduced the surface tension within the loop; the imbalance between surface tension inside and outside does the rest.

Droplets, Bubbles, Curved Interfaces

Tension in a wall produces pressure within and vice versa, as Laplace's law describes. So the pressure within a sphere such as a balloon should depend on the tension holding it together—more tension will produce more pressure. But the particular relationship depends on the size of the sphere. In a small sphere, tension will be more effective in producing pressure. Internal pressure will be wall tension divided by a measure of size such as diameter.

That wall tension can as easily result from surface tension as from a stretched elastic.† That, then, links the effectiveness of surface tension and the size of a droplet of water. The water needn't form a complete sphere; if it's only part of a sphere (such as a dome) we just use a mea-

---

*The value of the surface tension of pure water against air is around 0.072 newtons per meter at 25°C (77°F). It varies inversely with temperature, but far less severely than viscosity does, so a temperature correction is needed only for extremes or fine comparisons. Note that surface tension is a force per unit distance, not per unit area. That's because it happens not across a surface, like pressure, but along a line, the line of contact of air, water, and solid, as with a drop of water on a solid surface.

† We just have to replace the T of wall tension with the lowercase gamma, γ, of surface tension, so

$$\Delta p = \frac{2\gamma}{r}.$$

sure of its size as if it were complete. The usual measure is called the "radius of curvature," of the corresponding complete sphere. Within a hollow bubble, the pressure is doubled—air inside finds itself squeezed by both outside and inside interfaces, the ones between water and outside air and between water and inside air (or other gas).

What about a bubble of air in water instead of a droplet of water in air? The same rule applies. Intermolecular forces on the water molecules surrounding the air will pull inward, much like a circle of people holding hands who simultaneously pull on one another. The smaller the bubble, the more effectively surface tension will generate pressure. Similarly, the more effectively surface tension will resist an applied pressure.

That's the critical point to which all this verbiage has been leading. Surface tension rarely amounts to more than a mild nuisance on our corporeal scale—leave a wet piece of absorbent cotton out to dry, and it clumps up into a tight wad. Surface tension in the water between the fibers pulls ever inward, drawing the wetted fibers with its receding surface as the water evaporates. Hang a wet towel or T-shirt to dry, and either stiffens just a bit—it needs "fluffing up" so the surface fibers once again protrude properly. We minimize such problems with detergents, which reduce surface tension severalfold.

But at a microscopic scale, that mild nuisance becomes one of nature's major players. Consider the inward pressure that surface tension generates in three droplets of pure water or three bubbles surrounded by water. One is a millimeter across, flea size; one a hundredth of a millimeter, cell size; and one yet another hundredfold smaller, a tenth of a micrometer across, virus size. For the first droplet, the squeeze is a mere 290 pascals, three-thousandths of the sea-level atmospheric background. For the second, it's 29,000 pascals, three-tenths of an atmosphere, something that one ought not to ignore. For the third, it's a monumental 2.9 million pascals, almost 30 atmospheres. On this planet at least, that's such a big deal that atmospheric pressure can now be ignored instead.

Holes in the Leaves

Back to the leaves and the question that began this chapter. If water can evaporate out of the leaves, then those water columns whose negative pressures we found so impressive must be open at their tops. Why doesn't air doesn't get sucked in and fill those columns? What provides the crucial barrier that allows unimpeded evaporation of water but excludes air? Surface tension, unaided, does the job, managing to withstand pressures of tens of atmospheres. On a small scale, it's no mean force.

The stomata distributed (mainly) on the lower surfaces of leaves can't be the scene of the action. For one thing, they're just too big to sustain such negative pressures across an air-water interface. With typical wide-open widths of 20 micrometers (and thus effective radii of around 10 micrometers), they could sustain 14,400 pascals, about a seventh of an atmosphere, at most. Even if only the weight of the water column below pulled on the interfaces across the stomata—never mind flow and other pressure eaters—plants would be limited to about a meter and a half in height. Besides, that's not the site of the interface. Recall that the leaf has a lot of air inside, and it contacts both ends of a stomate as well as filling it.

The place where liquid meets gas, where sap meets air, doesn't correspond to any item of classic plant microstructure. No great surprise—the size of the interfaces lies below what we can normally see with a light microscope. Even if the interfaces were large enough to see with the naked eye, they don't stain with the dyes microscopists rely on to make structures visible. Surrounding big, sturdy plant cells (unlike puny, flimsy animal cells) are cell walls made largely of cellulose. That cellulose isn't a solid and continuous shell but instead forms a feltwork of fibers, with a little glue holding things together where fibers contact one another. The critical interfaces, nothing at all to look at, extend across the gaps between the fibers.

How wide are these gaps? About a hundredth of a micrometer, which is about ten times smaller than the best resolution of a top-notch light microscope of the kind, but of much better quality than you may have used in some biology course. Plugging that size (halving to get radius, of course) and water's surface tension into Laplace's law yields a tolerable pressure of 29 million pascals, or 290 atmospheres. That could support a column of water almost 3 kilometers, or 2 miles, high. And that's well above (although not vastly beyond) the most extreme negative pressure ever measured in a plant, minus 120 atmospheres. Surface tension does it all, and unlike the sap ascent business itself, the explanation has never been at all controversial.

Figure 7.4 gives a diagrammatic view of the analogous materials. With

FIGURE 7.4. We've used glued feltworks of fibers since at least the invention of papyrus. Here are three: on the left, a piece of lens-cleaning tissue on a background of construction paper; on the right, a kitchen scrubber.

increasing pressure differences, the interface where liquid meets gas curves ever more strongly inward—toward the water-filled cells. But at no negative pressure are the interfaces curved enough to round up into air bubbles small enough to pass into the system.[2] With far less extreme pressure differences, the same phenomenon can make a fine-weave fabric effectively waterproof. Air may pass through, water may pass through, but an air-water interface will get hung up. A wet pillowcase, grasped to form a crimp at its opening and with the opening held downward, can hold enough air to keep a person afloat without actively treading water. A dry one can't.

Making—and Not Making—Bubbles

We're not quite done looking at how plants take liquid water from soil and put gaseous water—water vapor—into the atmosphere. Nor are we done with surface tension. Several more big-ticket items have been left dangling. The next question, then: how can the system resist bubble formation throughout its columns of water? Won't even the tiniest bubble behave like an exploding bomb when sucked on by these extreme negative pressures? The event would instantly fill a vessel with vapor and put it out of action.

Obviously, nothing could be worse for one of these columns of water than a bubble of vaporized water. Their whole existence depends on the cohesion of water molecules in the liquid state. Lacking that group consciousness, molecules of gases (including water vapor, a perfectly ordinary gas) can't sustain negative pressures at all. So if a bubble forms anywhere, it should expand until it relieves all that tension in the liquid of its column. The walls of the tube, normally pulled inward, would recoil outward—nothing in our world achieves *perfect* rigidity—with the additional space occupied by water vapor and any other gases that boil out of the sap. The leaf above would then lose the use of that column as soon as it exhausts the sap that's above the embolism.

Fortunately, the same bit of surface tension-plus-geometry that kept the air out of the top of each column at least ameliorates the problem—Laplace's law can at least help, even if it can't completely cure. Consider an extremely tiny air bubble. To borrow from a previous example, imagine one with a diameter of a hundredth of a micrometer. Water surrounds the bubble, and the surface tension of that water puts a mighty squeeze on the bubble. It will feel a gigantic inward pressure due to Laplace's law and surface tension. Now the higher the pressure, the more of a gas will dissolve in a liquid—that direct proportionality goes by the name of Henry's law, after William Henry (1775–1836), the English chemist who discovered the rule. Henry's law describes the effervescence of bubbles when you uncap a carbonated beverage and drop its pressure to atmospheric—dissolved gas decides it prefers to stay dissolved no longer. Compress the bubble, and Henry's law can save the day.

So, any gas in a tiny bubble will be under enormous pressure—in both figurative and physical senses—to dissolve in the surrounding water or, if water vapor, to return to the liquid state. Worse (from the bubble's viewpoint), any such dissolution will reduce the size of the bubble and increase the pressure inside still further. So once the process begins, quick as wink the bubble will disappear without a trace. Tiny bubbles of vapor, then, have an understandable reluctance to persist. For keeping sap properly liquid, that's great.

But can trees and leaves count on Dr. Henry to do the job? Boyle's law, older and more famous (established by Robert Boyle in 1661), says that the volume of any gas changes inversely with the pressure squeezing it. Since negative pressure represents a reduction in squeeze, it should allow a bubble to enlarge. Consider what a difference a negative pressure of, say, 20 atmospheres in the water around the bubble would make. That negative pressure would greatly encourage bubble enlargement, opposing the tendency for surface tension to make it collapse. Fortunately, Mr. Boyle proves unequal to the challenge of Dr. Henry—but only if bubbles are small. Surface tension puts a tenth-of-a-micrometer bubble under a positive pressure of 290 atmospheres. Next to that, a negative pressure of 20 atmospheres fades into insignificance. In short, surface tension really puts the squeeze on very small bubbles, and a good thing, too.

Which raises a curious question, important whether we worry about cavitation and embolisms in plant vessels or about a particular everyday phenomenon. If any initially tiny bubble will immediately collapse from its own sky-high internal pressure, how can a bubble of dissolved gas or water vapor ever develop in the first place? Consider a liquid that happens to be overfilled (supersaturated) with dissolved gas or exists at such a low pressure (or high temperature) that its own vapor would like to boil out. Perhaps you should pour a glass of beer and contemplate the conundrum as it warms, asking how to reconcile this argument with the reality that's staring you in the eyes.

Small bubbles are indeed as reluctant to come into existence as they are to persist. In a liquid even a little distance away from any solid surface, bubbles almost never form. Instead, the liquid becomes supersaturated with dissolved gas: it holds more gas than it would if bubbles were permitted to form. Or it becomes superheated: it stays liquid beyond its normal boiling point, the point at which it would break out into a rash of bubbles if it could only get them started. In this sense, ascending sap is superheated. The pressure within it is so low that it would prefer to boil at any temperature it encounters.

To start either boiling or outgassing takes some sort of nucleation device. This could be a patch on a surface at which adhesion fails because it's hydrophobic rather than hydrophilic. Measurements of the tensile strength of water may really measure failure of such adhesion instead of true internal cohesion. That would largely explain why the highest values measured fall far short of values calculated from the molecular properties of water. Nucleation might happen at an imperfection such as a scratch in the surface, where the accident of geometry provides a kind of womb for embryonic bubbles that haven't yet reached viable size. Or, it might happen at the surface of some rough material that's been dropped into the liquid. In chemistry labs, these go by the name of boiling chips: rough-surfaced inert bits that keep boiling from being sudden and episodic. As well as having nooks and crannies, boiling chips start with air in their interstices. Emergence of a bubble most often leaves a little gas behind to initiate growth of the next bubble.

Sustaining supersaturation, superheating, or negative pressure, then, demands smooth surfaces to which liquid could adhere. And that's a critical part of what makes the sap conduits of plants special. While they don't look smooth inside to either the eye or the ordinary microscope, they manage to be sufficiently smooth at the smaller relevant size scale. And water adheres to these conduits—they're splendidly hydrophilic. Huge negative pressures, yes, but under circumstances highly inimical to bubble initiation.

Consider, again, that glass of beer. From fermentation (or some industrial circumvention) the beer has acquired a good charge of dissolved CO_2. The stuff stays dissolved as long as the beer is kept sufficiently cold and pressurized. Opening its bottle reduces the pressure on the beer, and it immediately becomes supersaturated with CO_2. Allowing the temperature to rise makes matters worse. So dissolved gas goes gaseous and bubbles out—if it can manage the trick. Notice that bubbles form on the walls of the glass, break loose, and rise. You may look long and hard, but you're most unlikely to see a bubble form in midstream. Notice, furthermore, that bubbles tend to form repeatedly at certain spots; some locations may even generate steady streams of bubbles. Our stainless-steel cookware behaves similarly: in gentle boiling, the bubbles repeatedly emerge from discrete spots.

What makes a particular spot a good bubbler? As figure 7.5 shows, an irregular surface provides a nice nursery for neonatal bubbles by sheltering them from surface tension's terrible squeeze. In such a "cozy corner" a bubble can grow to viable size and then, impelled by its own buoyancy, break loose and rise. And that, almost always, leaves just a little gas behind in some recess within the scratch or flaw to get the next bubble going.

FIGURE 7.5. Bubbles coming off a broken bit of a brick in a beaker of water at (or just slightly above) water's boiling point.

We're immersed in consequences. Much of the special sensation of drinking carbonated beverages comes from release of CO_2 in the mouth. So you don't want too much of the gas bubbling out beforehand—unless you like flat drinks. No kidding—the quality of the glassware can affect the taste of the drink. Drink champagne from fine crystal that's never been washed with any abrasive cleaner. And bite your tongue when you make nice noises about bubbly that you're served in plastic glasses: most plastics are severely hydrophobic. Never drink beer from a mug with an unglazed interior. (If someone pours your beer into such a container, notice the particularly big head on it. I bet you never guessed how relevant this book would be.) If you boil water for making tea or coffee in a brand-new kettle of stainless steel, you might notice that boiling begins abruptly and even a little violently—the water superheats, because bubbles can't find a home on its polished surface. A little initial rubbing of the bottom of the interior with an abrasive cleanser provides a permanent fix.

[Another digression. A small bubble in a glass of champagne faces a war between two opposing physical phenomena. The extra pressure inside the bubble due to surface tension urges its contents to go back into solution. The supersaturation of the newly depressurized and warming liquid asks dissolved gas to exsolve (odd but useful word) into the bubble. The enologists tell us that the initial size of a free bubble that will grow is about 20 micrometers, and that it may expand to about a millimeter as it ascends—not because of the drop in hydrostatic pressure but from further exsolution during its brief rise.]

One famous (in certain circles at least) device takes specific advantage of this reluctance of bubbles to form. Nuclear physicists need to detect the subatomic particles that their accelerators make. In 1952, Donald Glaser invented the "bubble chamber," a container of liquid that could be held just above its normal boiling point and thus be superheated and unstable. A sufficiently energetic particle passing through the liquid would nucleate a track of tiny bubbles, which a nuclear physicist (in the other sense) could then photograph. (Glaser received the 1960 Nobel Prize in physics for inventing the device.) Liquid hydrogen was commonly used in the chamber; the explosion of the Cambridge Electron Accelerator at Harvard University in the mid-'60s left no uncertainty about its hazardousness.

Keeping Bubbles Confined

The plumbing of plants doesn't consist of simple pipes running uninterruptedly between particular points in the roots and particular areas within individual leaves. As mentioned in chapter 6, the conduits vary a lot in length, with one or another sort of perforated plate dividing them into segments. These plates add a bit to the resistance of sap to upward flow. Why interrupt the conduits? Again, we invoke Laplace's law. If a bubble occurs in one segment, a plate of narrow holes or slits can keep the embolism from expanding and filling the entire system. That seems to be important when trunks freeze and dissolved gas comes out of the now-frozen sap—as we know from observing bubbles in ordinary ice cubes, gases are less soluble in ice than in cold water. At a cost of at most a 10 percent increase in resistance to upward flow, perforated plates allow a conduit to limit the spread of an embolism. If a bubble does penetrate a perforated plate, it will quite likely be broken up into smaller bubbles. Surface tension (once again!) plus their greater surface area relative to their volume makes these smaller bubbles more likely to collapse back into solution, to obligingly self-destruct.

A simple demonstration first done several centuries ago showed that the conduits also have side-to-side connections. One can saw a little over halfway through a trunk from one side and then saw a little over halfway through from the other, with the second cut made a little above or below the first. (Perhaps you shouldn't try this at home.) The cuts interrupt all directly upward paths for sap ascent. But somehow water moves around the cuts, and (if the tree doesn't fall over) sap continues to rise, and the leaves to transpire water.

[Still another digression. By contrast, if a cut encircles the trunk, a cut only deep enough to penetrate the actively dividing cells, the system breaks down, slowly but surely killing the portion of the tree above the cut. The technique, girdling, finds occasional use in forestry and ante-dates recorded history. If you want to clear a patch of forest to plant a crop, why go to the considerable trouble of cutting down the trees? If your tools are made of stone, that's a formidable task. And if you're not going to plow, it's unnecessary anyway. Just girdle trees during the winter, when they're leafless, and plant between the trunks in the spring. The trunks can even support climbing plants such as various legumes. The occasional branch that drops becomes firewood, as do the trunks when they finally topple. Eventually the stumps rot, providing a fully cleared field. The fields that the first European settlers found in New England— fields that allowed them to grow enough food their first summer—had most likely been produced just this way. They had been abandoned during the terrible epidemics suffered by the natives shortly before the Europeans' arrival, epidemics of Old World diseases introduced by itinerant sailors and fishermen.[3]]

Side-to-side movement of sap takes place through various kinds of small but numerous interconnections, pits visible but unremarked on in figure 6.6. Pits come in several types, as in figure 7.6, all with some kind of barrier to flow. In hardwoods, these may consist of little more than thin, porous regions of cell wall. These regions may be partly obscured by adjacent outgrowths of the main interconduit walls.[4] Pores within the pits are small enough so that Laplace's law (yet again!) restricts the passage of liquid-gas (or liquid-vacuum, nearly) interfaces. Or entire pits may operate as check valves, with central, movable occlusions that can block passage if the pressure difference exceeds some critical value. That's a common arrangement in softwoods. One way or another, these devices keep embolisms from spreading from conduit to parallel conduit.

Enough about what happens as sap goes up the trunk of a tree. Still,

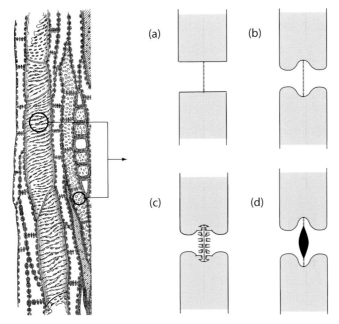

FIGURE 7.6. Water conduits (xylem) have long been known to have side-to-side connections—"pits." Contemporary microscopy shows that many have crosswise, microporous barriers, much like the wall of a typical plant cell. Pits, here in cross section, may be simple (*a*), bordered (*b*), vestured (*c*), or equipped with a lenticular check valve (*d*). (Vestures, outgrowths of the surrounding wall, seem to provide some mechanical support for the barrier in the face of a high pressure difference.) Yes, the drawing on the left is the same as the one in figure 6.6.

before passing on to events in the roots, I ought to mention two topics that remain imperfectly understood (nice euphemism, that) and are perhaps better left unplumbed in this minimally technical account. First, how are embolisms repaired? Trees clearly can reestablish broken water columns each spring, but how they do it isn't entirely clear, and not for want of effort by some clever and imaginative investigators.[5] And second, what makes maple sap rise in the spring? It's the basis of a large-scale commercial activity in northeastern North America, and the 40-to-1 boil-down of sap to make the stuff I pour on my morning waffles has generated all sorts of cultural correlates. (In Alaska, a 100-to-1 boil-down of birch sap makes a similarly tasty product.) Roots aren't necessary for the process, but what's happening in the trunks remains uncertain.[6]

Sucking on Soil

Ultimately, a tree pulls water out of soil. A tree can extract surprising amounts of liquid water from soil that to us feels dry. The air within even pretty dry soil ordinarily has a relative humidity close to 100 percent, but that's water vapor, not liquid water, so it's no help to the tree. Were the tree's vessels simply open at the bottom and immersed in a basin of liquid water, things would be simpler and easier. There would be no further cost in pressure beyond those imposed by gravity and upward flow. But trees grow in soils, complex combinations of solids, gas mixtures, and water. Their branching arrays of soil-encased roots provide support—to which we'll turn a few chapters hence—and, of at least equal importance, roots extract water from soil. Which turns out to be another tricky and multi-faceted game.

A fine literature of dirty books deals with the physics, chemistry, and (even) biology of soil, all giving lots of space to water content, water movement, and the issue of water extraction.[7] Everything varies, soil to soil, place to place, day to day: the relative contributions of solid, liquid, and gas phases; the chemical composition of each of the three (soil air can be quite different from the atmospheric air just above it); and the types, sizes, and size distributions of solid particles.

Roots have to contend with or co-opt forces from a variety of physical agencies. So we have another excuse for introducing elements of the physical world, both old ones and at least one that hasn't previously appeared in the present account.

First, gravitational pressure. Water prefers to run downward under gravity, which a root has to work against, but that's the least of its worries. Once in a while, gravity can even work for it rather than against it. Some soils swell when wet, and the downward weight of the solid stuff (denser, overall, than water) can force water upward. But that won't push water up much beyond ground level, so in the underworld gravity isn't a big deal either way, at least compared to its role aboveground. Few plants (and no big trees of which I'm aware) have roots that go downward nearly as far as their trunks go upward. Very tall trees commonly have startlingly shallow root systems, as in figure 7.7. So the pressure needed to get water to the surface against the pull of gravity doesn't amount to much.

FIGURE 7.7. The footplates of trees can be remarkably wide. Some, such as this one, are also notably stiff—here sufficiently stiff to have held up the trunk for several years before the picture was taken. The tree is a Douglas fir, on the grounds of the Friday Harbor Laboratory, San Juan Island, Washington.

Second, osmotic pressure. This came up in chapter 5 in connection with the guard cells that control the passage of gases in and out of leaves. We might image a scheme in which roots release some small molecules or ions into their passageways that draw water inward from the soil by osmosis. On the one hand, water in soil doesn't come uncontaminated, slightly reducing the efficacy of such a mechanism. On the other, and of greater importance, the sap that rises up a tree trunk has to be nearly free of dissolved material. So much water gets transpired that the accumulation of dissolved solids, coming out of solution as water evaporated in the leaves, would make big trouble as the growing season advanced. And dissolved gases could initiate embolisms in the vessels.

While osmotically generated root pressure does play a role, the main thing that draws water from soil is that severely negative pressure we've now been carrying on about for almost two chapters. The component expended in pulling water out of soil often goes by the name of matric pressure or matric potential.* It's really just another pressure loss, along with those of gravity and flow. The basic problem is that soil doesn't just contain water, it latches on to it, at times all too avidly. Some of the attraction represents chemical bonds, but most of it reduces to the same thing that kept air from entering leaves at the top of the system: surface tension and Laplace's law. Three principal conditions are relevant here.

1. Soils contain gases, liquids, and solids.
2. The solids, as fine particles, have an enormous overall surface area.
3. The particles are hydrophilic to a fault.

These particles of mineral may vary a lot in composition, but most are small, in the range of 1 to 10 micrometers, or thousandths of millimeters. They're also irregularly shaped. This combination explains their enormous aggregate surface areas—another place we invoke the rule that small objects have much more surface relative to their volumes than do large ones. A typical value might be 100 square meters of particle surface for each gram of soil. To put that in more familiar terms, a pound of soil represents about 11 acres of surface, something to contemplate when you lift a several pounds in a shovelful.

The hydrophilicity of soil should come as no great surprise after what was said earlier about glass, since for the most part it's made of similar siliceous stuff. That avidity for water has a lot of consequences we take for granted, quite beyond making mud pies practical. (Do children still make them, I wonder, in this preprogrammed, high-tech, sanitized age?) When mixed in the right concentration, water glues soil particles together as the water tries to minimize its contact with air. That facilitates making pot-

*The literature in plant physiology uses the more general terms *water potential*, *solute potential*, and *matric potential* where we're using the more familiar terms *pressure*, *osmotic pressure*, and *matric pressure*. The same literature often expresses them in units of energy per volume. Don't worry—everything boils down without specific conversions to pressures. Force per area is the same, dimensionally, as energy per volume.

tery, causing wet, newly thrown pots to harden rather than turning to powder before firing. It also helps to compact sundried bricks in just the same way that drying compresses a fluff of absorbent cotton, mentioned earlier.

The water in soil divides loosely into two sorts: gravitational water and capillary water. The first refers simply to water on which gravity acts to drive it downward enough to matter, meaning it occurs in relatively wet soil. In all but pretty sodden soil, almost all the water lies deep in the interstices between soil particles, and gravity can't get much of a grip on it. In those interstices, inwardly curving interfaces separate the second, capillary water, from the gas phase of the soil, as in figure 7.8.

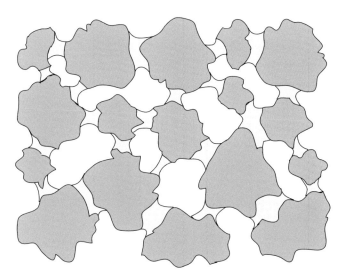

FIGURE 7.8. Air-water interfaces abound in any soil that's not powder dry or waterlogged. As soil dries (and sometimes shrinks), the interfaces retreat and become more concave, so a plant has to exert more and more suction to extract liquid water.

These interfaces, like any self-respecting air-water interfaces, want to minimize their areas, which they could do by sucking in additional water. A root, drawing water out of the crevices, has to make the opposite happen. In drawing water out from between hydrophilic particles, it has to create additional interface. So it finds itself working against surface tension, and Laplace's law (yet again!) describes that effort. The pressures it takes can be huge, especially where soil particles are tiny—these generate even smaller radii of curvature on their bits of interstitial water. Roots have to suck hard if water is at all in short supply.* They have to suck hardest in dry soil, where the remaining water lies especially deep in the crevices between particles, and where the interfacial radii of curvature are especially minute.

A lot of enemies oppose the leaf's efforts to draw up water: gravity due to the height of the tree, flow resistance due to the smallness of the conduits, osmotic resistance due to solutes in groundwater, matric resistance in the soil due to surface tension, and some lesser opponents. We might expect the most extreme negative pressures to occur in the vessels of the tallest trees, reflecting gravity and resistance to flow. But the matric pressure of the soil may be comparable to all the pressures talked about in this chapter and the last put together. The most extreme values occur in plants that extract water from the driest substrata, where matric resistance dominates. The record datum, I believe, was obtained by a former colleague, Bill Schlessinger, on a short desert shrub in the American Southwest—the minus 120 atmospheres mentioned earlier. As expected, most of that pressure came from the matric resistance of the soil.

Not that the osmotic opponent, salinity, can be dismissed. If the water in the soil contains a high concentration of dissolved material, then the system has to suck yet harder to make sap from soil water. If, for instance, soil contained water with a gram of sodium chloride dissolved in each kilogram (or liter) of liquid, the osmotic pressure opposing water extraction would come to a little less than 40,000 pascals, or 0.4 atmospheres. That's significant if not enormous—enough to hold up a column about 4 meters (13 feet) high and several times human blood pressure. In practice, the mineral content of natural waters varies widely. Soft lake water has a lot lower osmotic pressure, about 2500 pascals, or 0.025 atmospheres; hard river water has a higher osmotic pressure, about 60,000 pascals, or 0.6 atmospheres.

*How strongly do they pull? The interfaces must vary from spherical to cylindrical surfaces, so a combination (or statistical mix) of two earlier equations should apply. Thus

$$\Delta p = \left(\frac{\gamma}{r_1} + \frac{\gamma}{r_2} \right).$$

For a sphere, the two radii, the r's, are equal; for a cylinder, r_2 is infinite, so the second term is zero. For a cylindrical surface and an effective radius of 0.1 micrometers, the pull is about 0.7 MPa, seven atmospheres.

What about a plant that has only seawater from which to extract sap? Seawater has an osmotic pressure (against pure water) of over 2.5 million pascals, or 25 atmospheres. Dealing with that level of opposition takes mighty suction and some special cellular machinery. Should we be surprised that application of seawater to all but a few tolerant plants kills them PDQ? These tolerant ones—mainly salt-marsh grasses such as *Spartina* and the mangroves—don't exclude all the salt as they absorb water in their roots. A lot enters their system, gets raised to the leaves, and precipitated as water evaporates. If there's no precipitation to wash them, leaves may develop white incrustations of precipitate.[8]

Osmotic extraction costs energy, and even in salt-tolerant plants net photosynthetic productivity drops as the salinity of the surrounding water increases. Manipulating genes, much less ordinary selective breeding, hasn't yet produced many seawater-tolerant crops. Even a little salt in irrigation water slows water uptake by ordinary plants, reducing their growth rate and (as we judge a crop) their productivity. Worse, perhaps, salt remains behind in the soil, whether water evaporates from the ground or the plants extract it. Thus irrigating with even slightly saline water causes ever-increasing levels of salt in the soil—a problem that's now sufficiently widespread to measurably reduce the total amount of arable land on earth.

ON SOME WARM, windless day, grab a leaf that's exposed to full sunlight. It will be pleasantly warm to the touch. If our mammalian experience means anything, the conditions for photosynthesis ought to be just about ideal. The point I mean to push in this chapter is that our experience may mislead us.

The worst part of the problem is that we humans are a thick bunch—here in the ordinary thermal sense, not some insultingly mental one. Leaves are thin, much thinner than a hand. Since both hands and leaves are mostly water, a direct comparison is meaningful. Fingers are about 10 millimeters thick, while, for instance, a sun-exposed leaf of a large oak is only a fifth of a millimeter thick, fifty times thinner. When you grab a leaf, the second law of thermodynamics makes heat flow in whichever way will bring the temperatures of leaf and fingers closer together. A really hot leaf can't off-load enough heat to raise the temperature of the surface layers of your fingers much at all. Put another way, you cool the leaf more than a hot leaf heats you. The heat you gain equals the heat the leaf loses, but the temperature by which the leaf cools down greatly exceeds the temperature by which you heat up. So that pleasantly warm leaf was much hotter before you grabbed it.

On windless days, ordinary leaves on ordinary trees quite commonly run around 10°C, nearly 20°F, above the air that surrounds them. Under exceptional (exceptionally bad) circumstances they can reach twice that, meaning that on a hot day, say, at 35°C (95°F), a leaf might reach 55° (130°). That's about as hot as the hottest shower a normal person can briefly tolerate. On one hot afternoon recently, I measured 52.0°C (125.5°F) on a leaf of a white oak when the air temperature was 37.2°C (98.9°F)—the leaf was about 15°C (27°F) hotter. Such leaf temperatures are a lot hotter than the range within which photosynthesis works best: 20° to 35°C (68° to 95°F) for C_3 plants as are most trees. At temperatures above this range, too much metabolic output gets diverted into various respiratory processes. Fortunately, only occasionally do things become that bad, and then they are so only for the briefest periods—for most leaves.

8 Keeping Cool

Leaves near the ground, especially when sheltered from winds, must face an especially bad situation. I had brief use of an infrared imaging camera, a high-tech (= expensive) device that converts the long-wave emissions of leaves into false color. The particular fall day was clear but not nearly as calm as are typical midsummer days in the Carolina piedmont. Elevated leaves weren't all that warm, but the English ivy near the ground gave readings sometimes exceeding 100°F (38°C), so I focused (literally) on them, obtaining images such as those of figure 8.1.

What's more, such high temperatures happen despite a variety of devices with which a leaf keeps itself cool.[1] As chapter 2 noted, direct full sunlight is needlessly bright, since photosynthesis maxes out at lower intensity. After all the atmospheric absorption and after rejecting almost all the near-infrared portion of sunlight, a leaf absorbs about 1000 watts per square meter from an overhead sun shining through a clear sky. Photosynthesis consumes no more than perhaps 5 percent of that energy, an amount we'd dismiss as negligible were it not for life's total dependence on it. That other 95 percent makes trouble. Odd idea, admittedly—light as bad for leaves.

How to Keep Cool: Reradiation

Few of a leaf's problems run into such a dauntingly complicated mess of relevant physical processes as does overheating. Individual processes, the results of the processes, the interrelationships between the processes—straightforward describes none of these. But we can get a decent sense of what's going on if we first step through the basic issues and then weigh the various options that a plant can combine to avoid death-by-sunlight.[2]

If a leaf can be represented by a layer of water a fifth of a millimeter thick,[3] we can calculate how fast it should heat up under that 1000-watt energy input. The startling answer is that its temperature should go up by somewhat over a Celsius degree (about 2 Fahrenheit degrees) per second!* Sure, we ourselves heat up in the sun, as do our solar heaters (fortunately)

---

*1000 watts per square meter equals 1000 joules per square meter-second. Assuming the leaf is made of water, the leaf has a mass of 0.2 kilograms per square meter. Water has a heat capacity of 4200 joules per kilogram-kelvin—that's how much energy it takes to heat it by a degree. Putting these together ($1000/(200 \times 0.2)$), we get 1.2°C per second.

FIGURE 8.1. Two views of a leaf of English ivy. On the top is our familiar visual image; on the bottom, the same leaf as seen by an infrared-imaging camera. It converts the long-wavelength emissions of the leaf into false color, with red (ultimately white) the hottest and blue (ultimately black) as the coolest. The parts away from an edge or bent upward, like the lobes at the base, get hottest, while the edges and parts bent downward, the tip and outer lobes, stay coolest. On this particular day, air temperatures ran in the upper sixties (about 20°C) and leaf temperatures in the nineties (around 35°C). The camera, a very expensive FLIR SC620, was borrowed for a few hours.

> **Do it yourself:** With a handheld infrared thermometer such as the one mentioned in chapter 2, you, too, can explore the odd world of solar-heated leaves—plus the other fun and games that my grandchildren like to play with mine.

and our automobiles (unfortunately). But we and they have vastly more mass relative to the areas they expose to the sun. So heating proceeds at a far slower rate.

That rate of increase in temperature can't—and doesn't—continue. The hotter a leaf gets, the more effectively it off-loads energy, transferring it to its surroundings and, as a result, heating more slowly. Eventually it stops getting hotter altogether—heat output has now risen enough to equal heat input. How hot will a leaf become before reradiation to the sky becomes sufficient to stop further increase? That's also not too difficult to figure; it comes to about 84°C, around 180°F.* That's a little below water's boiling point and about the normal operating temperature of the coolant in automobiles before the advent of pressurized cooling systems.† In short, even assuming a cloudless, unpolluted sky, reradiation can't do the job, although it does its part. Indeed, it can't stop doing its part: in no simple way can radiant heat loss be stopped, even when the plant would be better off otherwise—as when reradiation at night drives leaf temperature below air temperature.

---

*We invoke the Stefan-Boltzmann equation, given in chapter 2, with a constant, again, of 5.67 watts per square meter-kelvin. A clear sky, the radiation sink, has an equivalent temperature of 220 K (–53°C, –63°F). Emissivities or absorptivities of leaf and sky for long-wave infrared (the wavelength as per the Wien equation, from that same chapter) are about 0.95. We just solve for the temperature that's far enough above that 220 K so that reradiation to the sky will reach 1000 watts per square meter. It comes to 357 K, or +84°C (183°F).

† Don't—I repeat, don't—ever remove the pressure cap from the radiator of a car that appears to be overheating—or even one that is working normally and has been recently driven. Removal of the cap causes the pressure to drop to 1 atmosphere and the boiling point of water to drop to 100°C, or 212°F. The pressurized liquid coolant, perhaps above that temperature, might stop behaving itself and come to an explosively sudden boil.

How to Keep Cool: Evaporation

A leaf may augment reradiation by calling on another physical agency, this one subject to control, adjustment, feedback, and other such good stuff. Water vaporizes only when supplied with sufficient energy, so evaporation cools leaves as it does our skins—unless the water fails to vaporize and appears as visible liquid perspiration. So transpiration cools a leaf. How much liquid water would leaves need to vaporize to drop their temperatures from the 84° just calculated to, say, a safely sublethal 44°C (111°F)? We start with the thickness of a typical leaf, the assumption, again, that it's made of water; the heat of vaporization of water in the relevant temperature range; and the heat capacity of water (as used earlier). Combining these, we can calculate that it would take a mere 15 milliliters of water (about half an ounce) per square meter of leaf area to do the job.* A drop in the bucket.

But that's a naïve way to look at what's going on. Sure, that small amount of water would bring down the temperature by the required amount. Sure, if continued sunlight weren't prolonging the agony. We need to view vaporization as a steady process rather than as a one-shot event. Say the leaf is faced with that steady 1000 watts per square meter and reradiates to the sky at the 44°C (111°F) assumed just above. How much of that 1000 watts is left for evaporation to off-load onto the atmosphere? The answer is about 330 watts per square meter.† Bear in mind that we're talking about energy per unit of time, not joules but watts, as well as per unit area.

How much water—per unit time—would be needed to transfer that amount of energy to the atmosphere by evaporation? Around 0.14 milliliters, half of a hundredth of an ounce, per square meter per second.‡

*Water has a heat of vaporization of about 2.43 megajoules per kilogram at atmospheric pressure and body temperature (37°C, 98.6°F)—the datum that matters for cooling by sweating. As the boiling temperature is approached, the heat of vaporization gradually drops, reaching 2.26 megajoules per kilogram at 100°C (212°F)—it's only slightly temperature dependent.

†One just uses the Stefan-Boltzmann equation, as in the top footnote on the previous page. For steady processes, heat capacity no longer matters—just heat of vaporization.

‡Getting that number takes no more than the heat of vaporization used in the top footnote above and gives 0.14 grams per square meter-second. One then converts from mass to volume: mass divided by density gives volume. Water's density is very nearly 1 kilogram per liter, or (in proper SI units) 1000 kilograms per cubic meter.

Which again sounds trivial—until you notice that we're talking about a small area, a square meter, and the briefest of times, a second. After proper extrapolation, things look anything but trivial. Making some educated guesses about periods of sunlight, growing season length, and so forth, and converting to the depth of a layer of water, I get a figure of around three-quarters of a meter. That's what a square meter of forest would have to suck out of its soil over a growing season just for keeping not exactly cool, merely not dangerously hot. Put in terms perhaps more familiar, that's a little less than 30 inches of water. It represents about the rainfall that would be received by typical temperate-zone forests—with nothing for direct evaporation from the soil, runoff, and other losses, not to mention droughts.

Bottom line: evaporative cooling can help, but it can't carry the entire load. And neither can reradiation. Furthermore, each has its odd disabilities. Reradiation requires a cold sky, so a sun that shines through an otherwise cloud-obscured sky presents a special risk. It also requires unobstructed exposure to that sky—partial blockage by adjacent leaves lessens its effectiveness, even if these aren't in the direct path of incoming sunlight. Beware of any shaft of light.

Besides taking water, often a precious resource, evaporation works poorly if the humidity is high. When leaf temperature exceeds air temperature, some evaporative cooling can occur even at a hundred percent humidity—the hundred percent refers to air temperature, not leaf temperature. Similarly, we can cool a bit by sweating when air temperature is below body temperature, whatever the relative humidity. (With air above body temperature, evaporative cooling can still work if the humidity is low—the change in physical state allows heat transfer from cooler to hotter body without thermodynamic lawlessness.) At least no matter how high the humidity, each unit of water evaporated off-loads the same amount of heat. The problem at high humidity comes instead from water's reluctance to evaporate.

How to Keep Cool: Convection

As you might guess from the persistent existence of leaves, the addition of one more physical agency ensures that heat output balances heat input at nonlethal temperatures. It's free, it's ordinary, but is it ever a complex business. I'm talking about convection, heat loss (or it could be gain) as air moves over a surface that's not at air temperature—like a lot of leaves, a lot of the time. Conduction through air works just like diffusion through air. It moves heat off the leaf surface and warms the air right next to a leaf. As a result, that air heats up. If the air rises all the way up to the temperature of the leaf's surface, heat transfer would then stop. But two things keep that from happening. Heat can be conducted from air right near the surface to air farther away. Fine, except that the thermal conductivity of air—or of any gas—wins no prizes. Beyond a trivial distance from the leaf, something else carries the load, that other thing being convection. So small a role is played by conduction in air that we can ignore it further—jumping from leaf to air may be critical, but what really sets performance is what happens to the warmed air next.

You'll notice, by the way, few explicit formulas in this section. Excellent physical rules handle reradiation and evaporation, prompting one footnote after another. Yes, formulas for convection do exist. But most are empirical, seat-of-the-pants encapsulations of experimental measurements and of limited value beyond the original experimental conditions. Others are dimensionless indices that mainly tell us what ballpark we're in.

Consider two kinds of electric space heaters offered for sale at your local hardware store. One kind has a heating element, a bar or coil, with perhaps a shiny reflective surface behind it and nothing more. Put it on a floor, and heat radiates from it, with heated air rising above it. The other kind adds a fan, which blows air forward, giving better circulation and permitting a more compact unit. But it produces both noise and wind—the last cools you by both convection and evaporation, so you have to turn its temperature up further to feel as warm. The first kind transfers heat by (besides radiation) what's called free convection. Here the driver is the buoyancy of warmed air. The difference in temperature between the hot air next to the heater element and the rest of the air produces a difference in density. The less dense warm air quite literally floats upward until it eventually mixes with the surrounding air, spreads across the ceiling, and causes a slow circulating motion in the air in the whole room.

With its fan, the other kind takes motion into its own hands, warming you faster, if in the end less pleasantly. It ignores buoyancy—which boils down to gravity—as a driver, so it will work in a spacecraft. This kind of heat transfer is called forced convection. The fan and motor may demand energy, but that, at least, costs nothing. The first law of thermodynamics tells us that energy is conserved, and the second law says that it all ends up as heat. So the work they do (at the expense of electricity) provides a little extra heat, and in no way less efficiently or effectively than what the heating element does. If all this heating happens in the great outdoors—to return to leaves—wind typically comes with the territory, so forced convection takes not only no extra energy but no physical fan.

Free convection. Except in spacecraft, heat rises. Since all gases expand when heated, it does so in every gas or gas mixture acted on by gravity. The situation is nearly as general in liquids: only water in the narrow range between 0°C and 4°C (32° to 39°F) goes the other way. That's something of great importance to life in lakes* but of no particular consequence here. How does that spontaneous, thermally driven rise depend on the particulars of the situation? That amounts to asking what drives free convection and what impedes it.

Buoyancy or, more explicitly, buoyant force drives the process. That depends on gravity first and foremost, and the more the better—even if it's not a controllable variable here on the surface of the earth. It depends, as well, on the difference in temperature between the fluid (here, air) right at the surface of the hot or cold solid and that of the fluid beyond the immediate thermal influence of that solid. And it depends on how much the fluid expands relative to a change in its temperature, what's called the thermal expansion coefficient. That's almost constant for gases at ordinary temperatures, almost as constant as gravity (gravitational acceleration, properly put). Opposing the process is viscosity, or viscous force. Fluids, we remind ourselves, don't take kindly to internal variations in speed, that is, to shearing motion. A buoyant updraft or downdraft represents just such shearing motion, and viscosity provides our measure of

---

*As lakes cool in the fall, the stratification established during the spring and summer, with cold water beneath warmer water, breaks down. When surface water drops toward 4°C (39°F), the temperature of maximum density, it begins to convect downward. That brings colder, and now less dense, water upward, restoring nutrients to the sun-illuminated shallows that had been depleted over the summer and leading to an autumnal bloom of phytoplankton and other microorganisms.

this antipathy. Air's viscosity at ordinary temperatures varies only a little more than does its thermal expansion coefficient.*

In the end, things subject to alteration are, first, that temperature difference—more is better for free convection—and size, where bigger is better. Envisioning the way the first matters is easy enough, but the second is more subtle. In effect, bigger means a larger convective plume, and that minimizes the pernicious effect of viscosity by reducing the steepness of the velocity gradients it dislikes.

The convective plume should be familiar to anyone who has noticed the distortion of the background behind a rising column of warm air—as above a nonsmoky fire or the burner on a stove. It was first illustrated by Robert Hooke, in the *Micrographia* of 1665, and it can be photographed with some optical tricks. (An analogous and fancier technique produces "schlieren" images, widely used for flow visualization in high-speed fluid dynamics.)

For our purposes, another visualization device gives more immediately relevant information. A form of liquid crystal material, available for decades, has the handy property of changing color with temperature. Better still, the temperature range over which it changes can be controlled. So we can buy temperature-indicating paint or coated sheets operating over several different ranges of temperature.[4] Figure 8.2 shows two rectangular pieces of such a sheet held horizontally and heated (and illuminated) by a lamp above them; except for free convection from the pieces, the air is still. The gradient in temperature from edge to center appears, one might say, in glorious color. No question about it, free convection does a lot better at cooling warm edges than cooling warm centers. The relevance of an upward-facing plate to a leaf on a tree should be obvious.

*The so-called Grashof number (Gr) provides a dimensionless index of the intensity of free convection; this evil-looking but revealing variable is the ratio of buoyant force to viscous force,

$$Gr = \frac{\rho^2 g \beta (\Delta T) l^3}{\mu^2},$$

where ρ, β, and μ are, respectively, the density, thermal expansion coefficient, and viscosity of the fluid, g is gravitational acceleration, ΔT is the temperature difference, and l is an average length of the surface exchanging heat with the fluid. For gases, β is $1/T$, where T is in kelvins—gases expand in direct proportion to the absolute temperature. At 30°C (86°F), that's 3.0×10^{-3} inverse degrees. More to come.

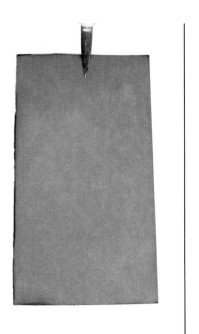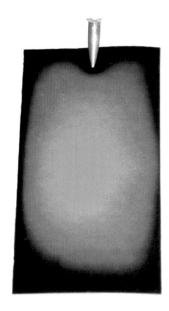

FIGURE 8.2. Two rectangular plates covered with liquid crystal temperature-indicating sheet, mounted horizontally and well away from a surface in a windless room. The left-hand one is of copper, whose thermal conductivity is especially high. The one on the right consists of cellulose acetate, with a thermal conductivity a thousand times less, not much different from that of a typical leaf. Both are being heated slightly above a room temperature of 21°C (70°F) by an overhead lamp. The sheet has a temperature range of 30°C (86°F) to 35°C (95°F), with its color running from black at the low end to blue at the high end of this range. The greater conductivity of the copper produces a much more uniform temperature: heat generated in the middle is more effectively conducted out toward the edges, where it can be most easily off-loaded to the air.

We need to distinguish between two different situations. If the plate consists of a material with a high thermal conductivity (fig. 8.2, left), heat received from the lamp can conduct outward from center to edge—moving down a temperature gradient, as thermodynamics insists. That makes the temperature more nearly uniform, so the hot center isn't quite so hot. And by making the edge hotter, it improves transfer to the air and lowers the average temperature. Good things, both—lower peak and lower average temperatures—if we think of a leaf illuminated by sunlight on a warm day. But leaves are nonmetallic, and so, unfortunately, they have low thermal conductivities, low enough so the edge-ward conductive heat transfer from the center is negligible, as on the right in figure 8.2.

The lower the conductivity, the higher both average and peak temperatures must be to achieve that convective heat loss. If conductivity across a plate or leaf is negligible, then the same amount of heat must come off every bit of the surface of the plate or leaf. After all, the lamp puts the same amount in, so when the temperature has stabilized, exactly as much must go out. What differs is the temperature needed to drive off that heat by free convection.

If the game consists of transferring solar heat from flat plate to air, a horizontal plate illuminated from above also happens to be a worst case. That orientation maximizes input; less obviously, it minimizes output. The trouble with output stems from what happens at the edge of the surface. Air heated from the lower surface flows upward around the edge, keeping it from being exposed anywhere to air at actual ambient temperature—in effect insulating it and reducing its convective effectiveness. As a result, for heat loss to balance heat gain, as it must in any steady-state situation, the leaf must be hotter if horizontal than in any other orientation.

What might a leaf do to improve its situation? A fix, at least a partial fix, consists of avoiding a square or circular shape. A more irregular outline provides not just more edge for transfer, but it means that on average, elements of leaf area will be closer to a free edge. That doesn't help through sideways heat conduction, still negligible, but it does reduce the buildup of semistagnant air on the lower surface and the upward flow of such air around the outer edges. The lower air temperature and steeper cross-flow speed gradient around the edges combine to cause a steeper temperature gradient. The steeper the temperature gradient, the more heat will be carried off. Still, while most leaves aren't square or circular, we shouldn't blame their odd shapes solely on this problem.

Some aspects of the shapes of some leaves on some trees, though, can be attributed to the problem of generating effective convection. Particularly revealing are the shape differences between leaves within individual trees that are exposed to direct sunlight and those ordinarily shaded. Admittedly, I say this with the passion not just of the persuaded but of the prophet. These differences between "sun leaves" and "shade leaves" defined my first episode of research on plants. Looking back, I think the project represented a reaction to my thesis work on flight in fruit flies, which left me anything but enamored with those awkwardly tiny creatures. Leaves took no culturing, they presented far less challenge to my dexterity, and they had only a little less personality.

In several of the larger oaks, most notably the white oak (*Quercus alba*), the sun leaves of an individual tree are much smaller and much more deeply scalloped than are the shade leaves. Bear in mind that we're not talking about genetics here—the full range occurs within a single individual, and the mix is probably readjusted in each year's new foliage.

In figure 8.3, sheets coated with encapsulated liquid crystals have been cut to the shapes of typical sun and shade leaves, and both kinds illuminated in otherwise still air—just as were the rectangular plates of figure 8.2. The sun leaf model doesn't become as hot as the shade leaf model,

FIGURE 8.3. Two model leaves made of poster board (another low-conductivity material) and covered with the same temperature-indicating sheet as in the previous figure. Both have been traced from leaves of a white oak tree: the one on the left, from a leaf well up in the tree and on the south-facing side; the one on the right, from a north-facing shady portion below. The deep indentations and limited area of the sun leaf make it a more effective heat dissipater, so under equivalent circumstances—horizontal orientation, still air, and radiant heating—it doesn't become as hot.

whether in terms of peak temperature or average surface temperature. These may be just models, but freshly picked real leaves give the same result. Maps of the isotherms from point-by-point measurements of temperature just don't produce such pretty pictures as do these models, and infrared imaging equipment wasn't available to me when I did the original work.

> **Do it yourself:** With a downwardly directed reflector-equipped light-bulb, you can try various shapes, both proper leaf shapes and any others that seem promising. Sheets of liquid-crystal-coated plastic (see endnote 4) are not especially expensive; ones with a temperature range of 30° to 35°C (86° to 95°F) or 35° to 40°C (95° to 104°F) work well for present purposes.

[Digression. A brand of frozen pizza with an unbaked crust ("rising crust pizza") is widely available in the United States. Instead of baking it on a pan for a few minutes, one is instructed to place it directly on the rack of a heated oven for about half an hour. I tried one and found that the edges overbaked and the middle remained somewhat soggy. So I devised a biomimetic version (fig. 8.4), which baked in about three-fourths of the normal time and came out considerably more uniform.

Convection improved—in this case, downward convection from a cold object in a hot atmosphere, like the rhododendron leaves of chapter 2. I initially intended to cut the frozen pie into an analog of the sun leaf of an oak, but I quickly thought of a better way, perhaps because frozen pizzas are hard to cut. Why slavishly copy nature, dependent on the severely constrained process of natural selection? Leaves grow outward, as we'll get to in chapter 13; pizzas do not. Putting a hole in the middle of the pizza is easy whatever the morphogenetic difficulty a tree might have in doing the same thing with its leaves. I wrote to the company, DiGiorno, pointing out the virtues of a toroidal, bagel-shaped pizza, but I received no answer.]

Forced convection and the mix. Pure free convection happens only under windless conditions. Otherwise we're talking about—or adding in—forced convection, which means wind, nothing more, nothing less. But things don't become blessedly simple with such a regime change (no more than they do in the Middle East). By the time forced convection is strong enough to dominate the situation, leaf temperature has shifted to something close to ambient temperature. Overheating then has ceased to be a

FIGURE 8.4. A practical implementation of the sun leaf's trick. Package directions for some frozen pizzas (such as DiGiorno's Rising Crust) call for their being baked, not on a pan, but directly on the rack of an oven. In practice, that overcooks the edges and undercooks the middle. Cutting a hole in the middle of the pizza vents cool air downward, just as the indentations in a leaf vent hot air upward. The pizza is unbaked in the top picture and baked at the bottom. (Making pizza from scratch beats any frozen version, so no product endorsement should be presumed.)

Do it yourself: I recommend the holey trick, if you must eat frozen pizzas, which, by contrast, I don't recommend. A hole around 5 centimeters (2 inches) across works well. Oh, yes—bake the cut-out piece as well. (Any pizza that bakes on a pan will derive little if any benefit from the scheme.) You'll never again wonder why doughnuts, bagels, and bundt cakes are toroidal.

particular problem. Heat transfer then mainly interests the people who work the account books, those concerned with environmental energy balances. Where the problem comes in, where the complexity is maximal, is in an all-too-common mixed regime: one dominated by neither free nor forced convection. Put another way, nature denies us a nicely simple and abrupt transition, as (usually) between laminar and turbulent flow.

Specifying, at least roughly, when the regime is mixed presents no great problem. I'll spare you the calculation, but will just put on the record that for a leaf about 5 centimeters across and 20°C (36°F) above air temperature, the regime is mixed between air speeds of 0.05 and 0.5 meters per second, between roughly a tenth and 1 mile per hour.* These speeds carry intuitive significance for almost no one, but we can at least tie the upper one to our perceptual world. The high (!) speed, 0.5 meters per second, is about a third of walking speed, and—significantly—it's about the minimum air movement that's ordinarily perceptible. So when the wind is so strong that merits being called a gentle breeze, forced convection dominates. It's effective enough so that leaf temperature and air temperature don't differ enough to matter. Where leaves have thermal trouble is in the unfamiliar world of imperceptibly slow air movement—we need a word other than *wind*, which carries quite the wrong connotation.

The low speed, 0.05 meters per second, is about as low as air ever moves in sunlit places for longer than a few seconds. For leaves, then, pure free convection must be rare. Nonetheless, it's real, a worst case, and living things must survive the worst that nature imposes on them. Similarly, the mix of free and forced convection must be important to a leaf at hazard of overheating. One might argue that in many places air never moves so slowly that free convection dominates. But it's hard to make that argument for a mixed regime. Mixed free and forced convection, though, has to be about the messiest of all the physical phenomena relevant to the lives of

*The index to the relative importance of free and forced convection compares the Grashof number (buoyant relative to viscous force), given in this chapter, with the Reynolds number (inertial relative to viscous force), given in chapter 6. Together they represent buoyant relative to inertial force. Specifically,

$$\frac{Gr}{Re^2} = \frac{g\beta(\Delta T)l}{v^2}.$$

If its value exceeds about 10, free convection predominates; if below 0.1, forced convection does.

leaves. Besides the basic analytic complications of the regime, the speed of air moving across a leaf in what we'd normally consider still air varies almost from second to second—according to the few measurements available. Furthermore, its direction varies as well as its speed. Free convection produces upward flow, and forced convection produces mainly sideways flow, so speed change means direction change as well. Worse yet, in nearly still air a tree as a whole can generate an upward pattern of free convection that individual leaves experience as upward forced convection.

As a result, on a calm day the temperature of a leaf on a tree that's exposed to direct sunlight fluctuates wildly. It may reach 20°C (36°F) above the temperature of the air during lulls, drop nearly to ambient temperature when the gentlest of breezes wafts across it, and drop fully back to ambient whenever a cloud obscures the sun. Figure 8.5 shows the tem-

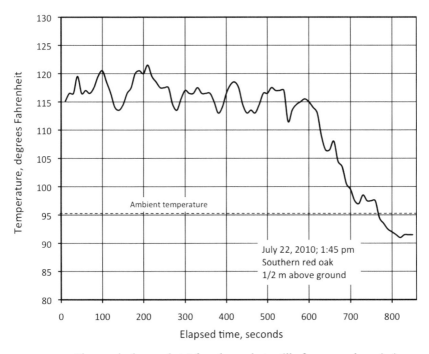

FIGURE 8.5. The graph shows what I found on a hot, still afternoon when aiming a temperature-indicating infrared gun (Raytek Minitemp MT6) at an oak leaf in a gap in the forest. Every ten seconds, I reported readings and anything else noteworthy to a tape recorder. The dips in the left-hand part of the graph correspond to slight movements of the leaf, movements generated by wind imperceptible to me. Between about 500 and 600 seconds, the leaf became shaded.

perature fluctuation of an ordinary leaf, here from a Southern red oak (*Quercus falcata*), over a few minutes. Except for the shading of clouds, which changes the input, the moment-to-moment variation comes from changes in the output, particularly the effectiveness of convective cooling. We have no idea how the biochemical processes within leaves contend with such wild temperature fluctuations. As a general rule, enzymatic reactions double or triple in speed with each 10°C (18°F) temperature increase, but individual enzymes behave, one might say, individually.

Besides putting a leaf's enzymes in an awkward situation, these moment-to-moment temperature changes will have another consequence, one that's unmentioned in the relevant literature, as far as I know. A 10°C (18°F) temperature change will cause about a 3 percent change in the volume of air within a leaf. If its stomata were shut tightly enough to seal the leaf (which is unlikely), the leaf would get around 1 percent thicker as the spongy layer expands. If the stomata are at all open, then a kind of thermal breathing would be unavoidable. Such in-and-out pumping means wind through the stomata, with who-knows-what effects on CO_2 ingress, water egress, even leaf temperature.

Not even the smallest terrestrial animals face the same problem. For one thing, all but the tiniest arthropods (insects, spiders, etc.) are thicker than a leaf, so heating and cooling take more time, and they respond more to longer-term (if still short) averages in environmental conditions. For another, most are mobile, capable of quick changes in orientation or location. A wasp or lizard warms up by seeking a sheltered place and exposing as much area to the sun as it can. It keeps from overheating by adopting a posture in which it casts the least possible shadow—meaning it intercepts the least possible sunlight. Or, either creature moves to shade or burrow.

Evasions

Not that a leaf doesn't have some tricks up its sleeve. The simplest consists of a less specialized version of what we saw in the models of white oak leaves. Most plants that experience direct sunlight and have limited access to water have relatively small leaves. Small size—recall the temperature-indicating sheets in figures 8.2 and 8.3—means more edge and less middle, which in turn means both lower peak and lower average temperatures. The leaves of desert plants that make leaves rarely grow large. Nor do the ones of plants that grow on well-drained sandy soil or coastal dunes. The extremes are pine needles, which make such good thermal contact with the air that they never deviate from air temperature by more than about a degree C (2°F). That reduced requirement for cooling in part rationalizes the much more limited ability of trees such as pines to move water up their trunks—recall their narrower conduits.

We've already seen how leaves reject the photosynthetically useless and thermally potent near-infrared component of sunlight. Many go further, rejecting a large fraction of visible wavelengths. Sometimes that's done by producing a hairy outer coating, a hair shirt for something other than the proverbial penance. Such leaves usually look light, even white, rather than dark green, with the light color coming from the reflection and scatter of their pubescence.

A particularly illuminating study compared two closely related species (in the genus *Encelia*) in which the one from a desert has hairy leaves and the one that lives in a moist environment has bald ones. (Esau versus Jacob, if you're into biblical allusions.) In the photosynthetically relevant spectral range, from 400 to 700 nanometers, the hairy ones absorb as much as 56 percent less sunlight. Interestingly, the pubescence doesn't appreciably reduce CO_2 entry, although it does reduce photosynthetic rate when the two species are tested under the same conditions in the laboratory. Of further interest, in the desert species newly formed leaves are smaller and hairier as the season becomes hotter and drier.[5]

Many cacti have a particularly neat version of the device. Cactus leaves have been reduced to spines, and cacti do their photosynthesis within their thick, fleshy stems. These cacti are green, but mainly when they're viewed from the side. Look at one from directly above, and it appears strikingly less green, sometimes even white—figure 8.6 shows the contrast. Such cacti carry umbrellas! Early and late in the day, their photosynthetic tissue looks out at the less intense oblique and more atmospherically filtered sun, while they avert their gaze, so to speak, from that dangerous overhead midday sun. They make a hairy coat do what we do with a membranous umbrella—just the way a lot of wind-dispersed seeds use a fluff of hair rather than a sheet of fabric as a parachute.

FIGURE 8.6. (*a*) A large cactus, with its umbrella on top. (*b*) A cholla, viewed from almost directly above. (*c*) Part of the same cholla, a section that apparently has been pushed over, so the sun strikes its side. Both cacti were photographed just after midday in November at the Atlanta Botanical Garden.

Again, for a broad leaf a horizontal orientation represents a worst case: maximum solar input and, especially for unlobed leaves, inefficient heat transfer in their output. Not holding leaves in a skyward-facing horizontal orientation is an obvious way to limit input. Plants have several versions of this tactic. Quite a few kinds of leaves shift away from being horizontal when the sun is strong and water isn't readily available—we say that they wilt, though the term carries a mildly pathological implication (figure 8.7*a* and *b*). Some leaves normally wilt at midday, whatever their access (usually limited) to liquid water.

My casual observations suggest that leaves with extensive lobing don't ordinarily wilt. That makes functional sense, since I found that lobing not

FIGURE 8.7. Leaves without indentations face a particularly severe heating problem on still, hot, dry days. Dogwood leaves, (a) and (b), react by wilting down to a more
nearly vertical orientation—intercepting less sunlight
and more effectively generating convective cooling currents. Blackjack oak leaves, (c), unusually large and unin-
dented for oaks, don't wilt. But, as in this side view, they
appear to avoid any near-horizontal orientation, giving
the tree a rather unattractively scruffy appearance.

only improves convective heat loss but also reduces the influence of orientation on such heat loss. Alternatively, some leaves just never orient themselves horizontally. Where I live, a species of oak (blackjack oak, *Q. marilandica*) with large, unlobed leaves that tolerates very well-drained soil does just that. Blackjack oak trees are sometimes described as "ragged" and look to me like models made by elementary school students for a play (fig. 8.7c).

A more subtle evasion should be familiar to every reader who has visited a florist's shop, a botanical garden, or a public greenhouse—even though the trick rarely draws comment. It takes advantage of two phenomena. First, winds near the surface of the earth vary from moment to moment, and those imperceptible ones vary at least as much as the breezes and gusts we notice. And second, an absorptive object doesn't heat instantly—the rate at which it heats depends on its mass and its heat capacity. Water has a high heat capacity, and water makes up most of the mass of a leaf. So the thicker the leaf, the more slowly it heats up when air movement drops. Of course to matter, the two phenomena, wind puffiness and leaf heating rate, must have similar time courses. What little data we have (very little, admittedly) tell us that yes, the time courses are indeed about the same. Other things being equal, then, the thicker the leaf, the lower the peak temperature it will reach during a lull—or near lull—in the movement of air across it.

The problem of heating during lulls will be worst for plants that can't turn on evaporative cooling when they get hot. Botanists have long recognized that the thickness of leaves correlates with the dryness of their habitats. Plants with those thick leaves go by the name succulents. That demands the immediate caution that no edibility is thereby implied. If anything, typical succulents come laden with noxious, indigestible, even poisonous contents. Of significance here: their extra water functions not as a reservoir for evaporative cooling but as a buffer against rapid heating, a device to deal with brief lulls in local air movement.

The talk so far has tacitly presumed that leaves are better off if they don't get warmed by sunlight. That may not always be so. Back in the second chapter, I referred to my casual measurements with an infrared thermometer of magnolia (*Magnolia grandiflora*) leaves on a cold, still day. Come summer, these big, elliptical leaves will be shaded; but in winter many have access to full sunlight, since the neighboring deciduous trees (sweetgum and maple, here) are bare. At an air temperature of –1°C (30°F), not a lot of photosynthesis will go on. But a leaf warmed to +18° (64°F) should be nicely poised to capitalize on that direct solar exposure. For understory trees, these magnolias do seem to grow rapidly. Holly (*Ilex opaca*) also grows in our understory and retains its leaves, which rise well above ambient temperature in the winter sunlight. The same thermal game almost certainly helps the spring flowers of our forest floor take photosynthetic advantage of early spring illumination, allowing them to do their business before the deciduous trees above leaf out.

Animal physiologists recognize a distinction between warm-blooded and cold-blooded animals, originally between, on the one hand, mammals and birds and, on the other, everything else. Nowadays we're more likely to paint the distinction as between endotherms, which keep warm by direct metabolic expenditure, and ectotherms, which stay at ambient temperature unless, for instance, they absorb sunlight, These broad, nondeciduous leaves suggest adding a third category or, better, dividing terrestrial organisms in yet another way. One group, we mammals and birds plus some other not-too-small creatures, hold body temperatures constant (and fairly high). A second group, the rest of the old cold-blooded animals plus needle-leafed plants, track ambient temperatures. The third have daily temperature excursions that exceed those of their surrounding air by a considerable margin. Extreme cases among this third group are those broad, nondeciduous leaves. Their temperatures may vary by over 40°C (70°F), between heat loss to the night sky and heat gain from daytime sunlight, easily twice the variation in ambient air temperature.

FALL HAS NOW TURNED to winter, and my annual raking again impresses me with the enormous area of leaf put out every year by a modest number of suburban trees. These relatively inefficient solar panels provide all the energy a tree can invest in growth, reproduction, and dispersal—the three central concerns of every organism that has ever lived. That takes real acreage: in a forest, aggregate leaf area greatly exceeds ground area.

Most broad leaves weigh very little: paper thin, a great pile of freshly fallen ones can be bagged and carried with little effort. An ordinary wheelbarrow, suitable for carrying dirt, holds all too little. A high-sided yard cart does better, and with such a light load one doesn't get back pains from its ergonomically awful crosswise handle. I once had the idea that the cost of operating a water heater could be reduced by running the input through a pile of composting leaves—the temperature of an active pile runs well above that of the surrounding air. A rough calculation disabused me of the idea, by showing that the elevated temperature came more from the pile's self-insulation than from actual heat generation, so low was its combustible mass.

Consider an odd implication of all that lightweight photosynthetic area. What if it rains on leaves when there's very little wind? Water is dense stuff, a bit denser than the leaves themselves. Wet leaves outweigh dry leaves, perhaps not as dramatically as a wet towel outweighs a dry one, but by appreciably more than their unwet weight. ("Dry weight" has another meaning in the botanical business, hence my use of the circumlocution "unwet.") So the sheer weight of soggy or water-coated leaves might present a structural challenge. In addition, a coating of water will hang around, and water provides a better biological medium than does air. It's likely to contain all sorts of spores and various microcreatures that lunch on leaves, ones that grow larger or more numerous over time.

On both accounts, weight and contamination, leaves ought to prefer that water run off them. Since few will be perfectly horizontal, especially when burdened with the extra weight of raindrops, that's most often what happens. Water runs off leaves like it does off the proverbial duck's back—and for the same reason. Both are quite hydrophobic. In all the talk about surface tension in chapter 7, nothing was said about the most obvious surface of all: the outside surface of a leaf.

How Hydrophobic Can You Get?

Surface tension, in the guise of capillarity, came up when we considered the height that water might rise in a glass tube or xylem conduit to which it feels some attraction. And surface tension came up again in explaining why air doesn't enter the open top ends of the water columns that run up trees. And it appeared yet again in the explanation of why a lot of negative pressure might be needed to draw water out of soil. In each of these instances, water has a high affinity for the local solid surface, whether clean glass, dirty soil (an oxymoron, I suppose), or the insides of the water-carrying conduits. Hydrophilicity, to put a word on it, is central. With no particular fanfare, we assumed perfect hydrophilicity, with water adhering splendidly to glass or leaf fibers or soil particles. Put in terms of a capillary tube, that adherence pulls the column of water straight upward, as on the left in figure 9.1.

The assumption of perfection got no close examination, because it describes reality well enough to need no hedging.

What matters here is the opposite phenomenon, hydrophobicity, where there's minimal adherence of water and solid surface. We're all familiar with it, even a little guilty at times because we don't encourage it sufficiently. I refer, of course, to the way water on a freshly waxed car beads up—along with most other common hydrocarbons, waxes are highly hydrophobic. "Water rolling off a duck's back"—the expression alludes to the oily coating of the duck's feathers, the waterproofing that keeps air trapped within its feathers and increases its buoyancy. In a capillary tube with a greasy inside wall (or made of a hydrophobic material), the water level will fall below the level of the water outside—capillary depression instead of capillary rise, as on the right in figure 9.1.

In practice, hydrophobicity doesn't approach the perfection of hydrophilicity. That's probably the reason we more often see the word *hydro-*

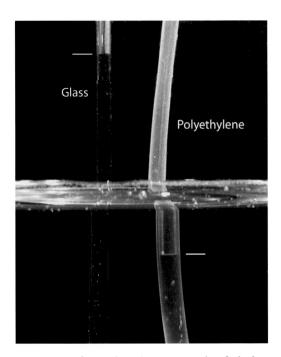

FIGURE 9.1. The meniscus in a narrow tube of a hydrophobic material drops below the water line, while the one
in a hydrophilic tube rises. In the first (*right*), repulsion of
water from the tube's wall augments gravity and impels
the water downward; in the second (*left*), attraction partly
offsets gravity, so the water rises.

phobic than the word *hydrophilic*—we worry more about the degree of the
first, about how hydrophobic some surface is. So we're now faced with a
range of performance rather than with some either-or business.* A few

*Something called contact angle provides the usual measure. The contact angle
is the angle between a solid substratum and the edge of a bit of liquid above it, as in
figure 9.2.

The relevant formulas use the cosine of that angle, as in the full form of the one for
capillary rise in the last chapter. Thus

$$h = \frac{2\gamma}{\rho g r} \text{ should really be } h = \frac{2\gamma \cos\Theta}{\rho g r},$$

where Θ is the contact angle. With nearly perfect wetting, as for water on clean glass,
Θ is very close to zero, so $\cos\Theta$ is about 1.0—and so we could ignore it. For water on
wax (as on a car), Θ is about 107°, and $\cos\Theta$ is –0.29. That makes h properly negative—

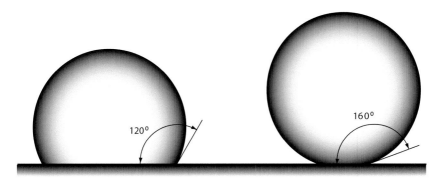

FIGURE 9.2. The more hydrophobic the surface, the greater the contact angle (defined here) between water droplet and surface—greater repulsion means that less air–water-surface contact line is needed to support the weight of the droplet. (Real droplets aren't quite spherical—gravity makes them sag a little, but that's a complication we don't need at the moment.)

years ago, that range took center stage, following casual observations that water rolled off some leaves a lot better than it rolled off the back of any ordinary duck. The leaves of the sacred lotus, *Nelumbo nucifera*, provided the first case that drew attention, so the effect, termed superhydrophobicity, also carries the less phobic name of lotus effect.*

What was going on? Lotus leaves turn out to be doing something for which textbook chapters on surface tension made no provision. Their trick depends on no great chemical novelty but instead on combining a normal hydrophobic waxy coating with a microtextured surface. As a result, droplets of water on these leaves became more nearly spherical. Many other leaves are also superhydrophobic: for instance, myrtle or donkey-tail splurge (*Euphorbia myrsinites*), a water fern (*Salvinia oblongifolia*), nasturtium (*Tropaeolum majus*), a eucalyptus (*E. macrocarpa*), blue seakale (*Crambe maritima*), and cabbage and its conspecifics (*Brassica oleracea*).[1] All these are plants commonly cultivated by people, meaning that

water level falls in a waxed capillary tube. We sometimes coat normally hydrophilic fibers with something that raises the contact angle in order to make a fabric water repellent.

As an exercise, you might use one of the formulas above to determine the magnification of the photograph in figure 9.1.

*Superhydrophobicity, combining physical and chemical surface treatment, can push Θ up to around 160°, not far short of the 180° that would represent perfection. The cosine of 160° is –0.94, within about 6 percent of the ideal, –1.0.

they're the kinds most easily available (and preidentified in greenhouses and botanical gardens) for people concerned with water shedding. So we're certainly looking at something far from rare, something you can explore yourself.

These plants don't either fall into any sharply defined botanical grouping or represent the inhabitants of any especially distinctive habitat. Nor, it seems, do plants hold the patent on the device. A variety of insects have superhydrophobic feet, most notably—and significantly—ones such as water striders, which run around on the surfaces of bodies of water while supported by surface tension. Some insects have superhydrophobic wings as well.

One might ask who discovered superhydrophobicity. As with many novelties in science, perhaps most of them, we too quickly rush to anoint a single hero. Here in particular we're looking at something closer to an awakening awareness. That rough or porous surfaces sometimes shed water better than do smooth surfaces with the exact same chemical composition was pointed out well over half a century ago.[2] But the difference attracted little attention until in the early '90s, when two botanists at Bonn, Barthlott and Neinhuis, took a careful instead of a casual look at droplets rolling around on lotus leaves.

Superhydrophobicity confers a benefit we might not have guessed. Not only do droplets of water roll off more readily. As they roll along on the surface of a leaf, they pick up and take unto themselves all manner of tiny particles. Most of these, perhaps all, are items the leaf is better off without: dirt, spores, and so forth. I saw lotus apparently thriving around a lake in Beijing. The air pollution from anthropogenic particulates completely obscured the midday sun every day except one during my two-week visit—typical for the city, I gather, except when hosting Olympic competition. Perhaps the combination of superhydrophobicity and the occasional rain gave these lotus leaves a leg up. I tested their trick by tossing some droplets on leaves in several formal gardens (when no one was looking); one casual trial produced figure 9.3.

Unless the optics of droplet-as-lens makes the pictures deceptive, the effect isn't at all subtle. As put by Barthlott and Neinhuis, "We observed a peculiar effect. Independently of the degree of pollution of the collection site, species with smooth leaf surfaces always had to be cleaned before ex-

FIGURE 9.3. Droplets of water (acting like little magnifying lenses) on a lotus leaf. Leaf surfaces just above veins appear to have slightly less hydrophobicity. What makes them more hospitable to water may just be their slight concavity. Photograph taken at the botanical garden in Sydney, Australia.

amination [with a microscope], while those with epicuticular wax crystals were almost completely free of contamination. . . . Based on a number of experiments, it was proven that water repellency causes an almost complete surface purification (self-cleaning effect): contaminating particles are picked up by water droplets or they adhere to the surface of the droplets and are then removed with the droplets as they roll off the leaves."[3]

Surface tension pulls—provides tension—along a surface, not in or out of a surface as does pressure. That explains something we've all observed but rarely if ever think about. One has to tip a hydrophobic surface (perhaps a sheet of plastic or some greasy glass) beyond a certain minimal angle before droplets will roll. Put a droplet on an inclined surface, as in figure 9.4, and the downhill side of the droplet makes a steeper angle with the surface than does the uphill side.

Consequently, the downhill side pulls back uphill more effectively than the uphill side pulls downhill, opposing gravity's urging that a droplet roll downward. When will the droplet begin to roll? It will start when

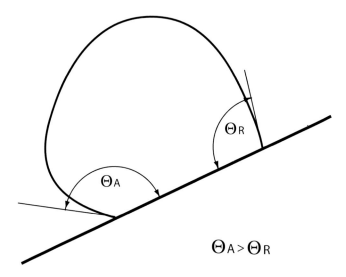

ΘA

ΘR

$\Theta A > \Theta R$

FIGURE 9.4. Gravity works on a droplet on an incline in an odd way. The contact angle on the downhill side exceeds that on the uphill side.

the surface is tipped enough so the gravitational force that draws us all earthward becomes greater than this asymmetry of force from surface tension.* We're accustomed to the difference between static and sliding friction, the way making some solid object start moving (against static friction) is harder than keeping it moving (against sliding friction). So we're likely to assume that nothing different is going on with rolling droplets. Beyond their gravitational motivation, though, solids sliding and droplets rolling aren't analogous phenomena.

That particle pickup predilection and other likely advantages have triggered a frenzy of biomimetic work. Just bounce *superhydrophobicity* off your favorite Internet search engine—technological contributions now vastly outnumber biological ones. Making superhydrophobic surfaces presents no insuperable challenge; many different approaches are

---

*Put with more formality, the lower or advancing contact angle is greater than the upper or receding contact angle. A difference of around 30° is typical. Every ratio of forces generates a dimensionless index, it seems. This one, gravitational to surface tension forces, $\rho g l^2 / \gamma$, goes by the name of Bond number. A low enough value (l is wetted perimeter) permits a creature to walk on water; unfortunately for those of us with human feet, that requires a 5-gram (⅕-ounce) walker.

possible for effectively combining chemistry and morphology, and a surprisingly wide range of advantages can be obtained. Imagine fabrics that are more water repellent but still porous, and painted surfaces that are cleaned by ordinary rain. A commercial coating, Lotusan, has appeared, although it's not without its problems—as typical for a first-generation product. At this point we've just (one might say) scratched the surface.

Superhydrophobicity might matter to leaves in another way as well, one perhaps parenthetical for our paradigmatic leaf-on-tree. What if water covered an entire leaf? When submerged, many leaves have a mirrorlike shine. They trap an air layer, and reflection from the fairly flat interface between that air and the surrounding water is what produces the shine. Many—but not all—wetland plants retain such gas films on their surfaces. If you have a nasturtium plant that's willing to donate a leaf, you can see the effect for yourself. Does that film on a fully submerged leaf have some utility, or is it nothing more than an accident of leaf surface chemistry?

A recent investigation has uncovered an important functional role.[4] A submerged leaf without a film must exchange oxygen and CO_2 either through the general cuticular surface or through the stomates. But the cuticle presents a strong barrier to diffusion, and the stomates cover only a tiny area. When they're dissolved in water, both gases diffuse far more slowly than they do in air. An adhering gas film greatly increases the area available for diffusion from the surrounding water into a gas, where diffusion works then much better. That provides what amounts to an external lung, which allows the stomata to operate in their normal fashion: passing molecules in and out by diffusion in the gas phase.

The function of this general device, a coating of gas on a submerged organism, first drew recognition in aquatic insects, in 1947.[5] The coating, called a plastron, fills the spaces among hydrophobic hairs on their general hydrophobic surfaces. The plastron provides lots of surface through which these air breathers can exchange respiratory gases with the surrounding water. Like the submerged leaves, they have a conspicuous silvery coating. The genus name of an aquatic spider, *Argyronectes*, alludes (*argyro* meaning "silvery") to that appearance. (Speaking of names, plastron is an unfortunate choice, since it's also used for the ventral shells of turtles, which serve no analogous function.)

Tipping and Dripping

You've undoubtedly observed, even if you haven't taken notice, that many leaves have pointed ends. Some have many such ends, though more often each has only one. There's little obvious correlation between sharp-tipped leaves and any aspect of a tree's growth form or habitat. For instance, where I live, the white oaks have rounded ends on their leaf lobes, while the red oaks have sharp points. Any functional significance, let alone any adaptive significance (bringing evolutionary advantage into the picture), remains obscure. Even with my long-standing adaptationist bias, I have to admit that the difference may be inconsequential. The general issue certainly hasn't prompted any attention that I've heard about from the botanical community.

One particular aspect of the leaf tip story, though, has drawn both observational and experimental scrutiny. Some leaves have highly exaggerated pointy tips. In the jargon of identification manuals, they're referred to as "acuminate," as opposed to merely "acute" for ones that taper to a point—figure 9.5 shows several. What has attracted attention is the peculiar prevalence of these forms in tropical rain forests, particularly among the leaves found in the understory.

So what function might they serve? The hoary old observation that leaves with these "drip tips" shed water more effectively has gained considerable support. (On the chance that I might shoot down an urban legend, I once tested it. Spraying water on leaves with and deprived of their drip tips while weighing them only confirmed the standard story.) The name remains appropriate—it doesn't lead teachers and textbook authors astray. But that pushes the issue down a level, posing the questions of why better water shedding should matter and why it should matter most in a rain forest. Complicating the picture, drip tips are neither a universal characteristic of these plants nor entirely limited to plants in rain forests. After at least a century of speculation, observation, and experimentation, these deeper questions remain without definitive answers.

One suggestion has these tips reducing the soil erosion caused by the splashing of drops of water as they strike the ground.[6] Leaves with drip tips do drop smaller (and more frequent) droplets: in one study the volumes of their droplets ranged between 8 and 30 cubic millimeters, contrasting with ones 50 to 150 cubic millimeters from leaves without drip

FIGURE 9.5. Drip tips on several leaves from distantly related plants in our departmental greenhouse.

tips.* But the postulated effect on the soil remains hypothetical. Even if the effect were established, adaptive significance—survival and reproductive value—for a tree doesn't necessarily follow.

A substantial literature looks at whether or not (and of course to what degree) better shedding of water reduces colonization of a leaf's surface by epiphytic lichens, fungi, and mosses. Investigators have compared plants with different degrees of tip elongation that live near one another. They have watched colonization of mounted artificial leaves of varying shapes. And they've altered real leaves, trimming back the elongate tips or (as sham operations) done other bits of reshaping. Leaf surfaces certainly dry more rapidly if the tips have that peculiar long, tapered shape. And epiphytic colonization can reduce light interception by half or more,

*I was taught that it took 20 drops from an eye dropper to make 1 milliliter, or 1 cubic centimeter. That's 50 cubic millimeters per drop. A drop, though, constitutes no standard quantity, with the value depending on the density, viscosity, and surface tension of the liquid, plus the geometry and hydrophobicity of the tip of the dropper.

so avoiding it when you live in a shady understory might matter. But whether a drip tip reduces long-term colonization remains problematic. Some studies show a significant (if small) effect; some fail to find anything that withstands statistical scrutiny.

As usual, I can't write or think about any suggested device without both wondering about the evidence supporting it and making my own guesses about its functional significance. Half a century of doing experimental work on how organisms function does that to a person. So here are some suggestions that don't seem to have been considered—with apologies to both reader and investigator if I've missed something in the published literature.

First, why drip tips? As suggested earlier in the chapter, maybe what bothers the leaf is the sheer weight of water on it. Leaves are thin and less dense than water, and they stick out from twigs on slender leaf stems—their petioles. Particularly in the understory, and especially in the particularly dark understory of a rain forest, it should be important to maintain a proper orientation, one that maximizes exposure to light. Water that hangs around will deflect a leaf downward, and it will do so with particular effectiveness if it's out near the far end of the leaf. One study found that a significantly greater amount of water hung onto leaves without drip tips between a half hour and an hour after a rain, and the investigator noted (without comment) some downward deflection of the leaf surface.[7]

And second, why are drip tips so much less common in dryer and colder climates? Dryer? The greater weight (per area) of leaves that occur in dry places as well as the greater light intensity in any understory might explain the scarcity of drip tips in dryer places. A little water would make less difference to overall weight, and a little more inclination might not matter where sunlight abounds. Colder? I wonder whether drip tips present a peculiar problem. Once in a while, a combination of gentle rain, low wind, and a freezing temperature near the ground will cause icing on exposed surfaces. The ice can approach a centimeter in thickness, so branches break off some trees, other trees snap or uproot, and general mayhem prevails. While icing most often happens when deciduous trees lack leaves, it does sometimes occur when they're fully leafed out—and then with particular destructiveness.

A tree can tolerate full defoliation once in a while, but uprooting and major branch losses are worse hazards to survival. Such events, even if rare, take on especial significance if they pose an existential threat to a long-lived organism. The hazard is greatest if that organism does not reproduce until well on to maturity. That's a common gambit among big trees: invest in growth to capture a place in the sun before diverting resources to making acorns.

Ice is a lousy solid. It combines stiffness and brittleness to a greater degree than metals, plastics, or biological materials—which is to say that cracks propagate through it with ease. An ice crusher is much easier to design and takes a lot less power to run than does a stone crusher. And even stones, including our concocted concrete, crack fairly easily. If we want to build a decently crack-resistant structure, we embed a lot of steel rod (called rebar) in the concrete. Look at pictures of concrete buildings that have collapsed in earthquakes, and you'll see how little metal was incorporated during construction. (Since no sign of the rip-off will be visible when the building is completed, shorting the steel must be a great temptation. That's one reason building inspectors make repeated visits to construction sites.) Icicles crack off with very little provocation. Unless—and here's my guess—they have some fibrous material holding them together, material that withstands tension just like the rebar in a concrete building. That's the basis of an exotic material called pykrete, ice with a lot of wood pulp inside. So maybe a drip tip allows icicles to grow long and heavy, keeping them attached even when they crack crosswise. That might allow them to get heavy enough to increase a tree's risk of life-endangering failure.[8]

The idea should be readily testable, at least for the crude initial check that we should always seek before investing a lot of time, energy, and money in a rigorous test of some odd hypothesis. A pilot experiment should take nothing more than a variety of expendable leaves—many ornamentals have drip tips—and a walk-in cold room or even a chest freezer.

Drip tips might provide practical guidance for the design of teapot spouts, where we want (among other things) to avoid a clinging drop that might stop clinging in some inappropriate place. One can buy a tiny metal insert for a spout that claims to minimize the problem of liquid either clinging or running down the outside of the spout. It works fairly well, but the particular prosthesis can be hard to find. That its designer had the drip tips of leaves in mind seems quite unlikely. But that's the fly in the

> **Several do-it-yourself projects:** (1) If sawdust of any coarseness is available, it's easy to make pykrete by packing the sawdust into a plastic ice cream container, filling the container with water, and freezing it. For a dramatic demonstration, hammer nails into the resulting material. (2) Impressively long icicles can be made by hanging long strings from points where cold water drips. The strings keep fractures from parting, allowing further drips to heal them. (3) Try your own drip tips if you have a refrigerator with a side freezer, or use a separate freezer. You might make models out of acetate sheet, which is fairly hydrophobic, or modify real leaves. A spray gun or dripper emerging from a container of water with some ice cubes in it to fix its temperature ought to give adequate reproducibility. I'd use our digital kitchen scale, but for a first pass that's really just a flourish. (I admit that just about anything looks simple before one actually gives it a try, and I haven't tried this one yet.)

ointment for most purported instances of biomimetics—we usually notice the natural analog only after we've made something ourselves.

Multifunctionality: Shedding Spores and Such

Leaves, to repeat, can offer attractive habitats for smaller organisms. They're wet, at least modestly nutritious, and well positioned for the launch of airborne propagules—the generic term for seeds, spores, and so forth. Their defense? Chemistry can do the trick, at least until a pathogen evolves resistance. Surface protection may help, and we'll return to that in the next section. The simplest first line of defense can simply be a surface smooth enough so that the same wind that deposited a spore (or else a subsequent breeze) blows it off again. Leaves appear to play this game, despite a subtle aerodynamic problem.

Back in chapter 4, we considered a peculiarity of the way air or water flowed across a surface. At the very surface, nothing flowed—the velocity was zero. As distance outward from the surface increased, the speed of flow increased, rapidly at first and then less rapidly as it gradually approached the speed of the otherwise undisturbed air or water. Thus even in a substantial breeze, wind speeds right down near a surface are quite low. Nonetheless, a spore that fails to find a crevice cannot avoid some wind, wind that might roll or push it off the leaf, as in figure 9.6. The

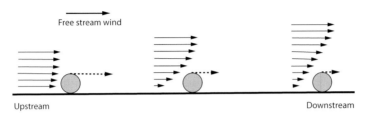

FIGURE 9.6. The farther a particle such as a spore lies from the windward edge of a leaf, the lower the wind that strikes it—other things such as the speed of the overall wind being constant. Lengths of solid arrows indicate local wind speeds.

broader the leaf, the worse the problem will be, since the region of retarded flow gets thicker with the distance downstream from the edge struck by the wind. Put more pointedly, the speed gradient above the zero-speed surface becomes less steep with distance from the upwind edge,[9] so a spore sitting on the surface will feel a smaller fraction of the overall wind.*

One partial evasion of the problem takes advantage of the puffiness of natural winds near the earth's surface. The gradient region take just a little time to mature, so suddenly gusting breezes are more effective in blowing off particles than are steady air movements.[10]

Since spores ultimately land on the surface as a result of gravity, the upper surface should be at greatest hazard. So a smooth upper surface should be an advantage, more of an advantage than a smooth lower sur-

* A formula, unfortunately with various limiting assumptions, relates overall flow (v), distance downstream from the edge (x), and distance outward from the surface (z) to the local downstream flow (v_{local}) that might be felt by a spore:

$$v_{local} = 0.32 v z \sqrt{\frac{\rho v}{\mu x}},$$

where ρ and μ are the density and viscosity of the fluid.

face. We don't ask why veins protrude beneath rather than above a leaf, perhaps because we build roofs with their supporting rafters beneath. But neither of our rationales, ease of waterproofing or our preference for joints that push (as roofing against supporting rafters) rather than pull, should matter for leaves. As for veins on the leaves' undersides, minimizing spore-sheltering crevices isn't likely to be the whole story, but it might be a factor.

More Multifunctionality: Hairy Surfaces

Talking about the surfaces of leaves brings up a more general issue. How can we assert that some feature serves some significant function in the life of an organism? And if we do make a claim of function, how can we further assert (or know enough not to assert further) that the function happens to be the particular one that has driven the evolution of a feature?

Like adult human males, some leaves have hair and others do not. Hairy—or fuzzy or pubescent—leaves form no tight phylogenetic cluster that might indicate common ancestry. So we're looking at a convergence, meaning that the most recent common ancestor lacked the feature, that hairiness is a shared, nonderived feature. People who care about the evolutionary interrelationships of organisms look for shared, derived features, indicative of common ancestry. People (like me) more concerned with how things work look (or should look) for these shared, nonderived features. The convergences they represent are fingers that often point to functional significance. After all, functional advantage—leading as it will to greater reproductive success—is the fuel for natural selection.

Chapter 8 pointed out that a hairy surface could reflect and scatter light, and that this reduction in sunlight absorbed could keep a leaf from getting too hot or expending as much water in keeping itself from getting too hot. The suggestion satisfies by rationalizing a common but nonuniversal feature, the occurrence of hairy leaves in relatively sunny, dry places. With ample water or sustained shade, why bother? If the habitat has some minimum level of daytime air movement or if a leaf's enzymes have sufficient thermal tolerance, again, why bother? So the iffiness of occurrence of the feature doesn't seem strange. But the fit of habitat character and occurrence of the feature is more like bathrobe than pantyhose.

The assertion of function on the basis of convergence under common environmental conditions has to avoid a lurking pitfall. What if the feature serves more than one function? We might be able to figure out, by looking more closely at who lives where, what the primary—in terms of natural selection—function might be. Still, what if the feature serves one function under one set of environmental circumstances and another function under another set?

Hairy leaves provide an excellent example of just this problem. Sure, I like that thermal function. But I bear in mind the old adage that if your tool is a hammer, all the problems look like nails. And I worked on convective cooling in leaves. Here, then, are some functions for leaf pubescence for which either evidence exists or a reasonable surmise can be made:[11]

- Reflection of excess light that would otherwise be scattered onto the main photosynthetic tissue, as already mentioned.
- Reduction of water loss by increasing the distance water vapor has to diffuse before it can be swept away by moving air. This increases the humidity just outside the stomata.
- Protection against excessive heat loss by convection limitation. This might keep leaf temperature above disfunctionally cold air temperature, closely analogous to the main function of animal fur.
- Protection against excessive heat loss by limiting reradiation from a leaf's surface, particularly at night. This means letting the more cold-tolerant and photosynthetically irrelevant hairs do the reradiation.
- Protection from herbivores, both insects and vertebrates, by provision of detachable bits that are irritating, noxious, or seriously poisonous. I itch whenever I contact the leaves of our fig tree, and that's a mild bother as surface irritants and poisons go.
- Protection from the spores of airborne pathogenic organisms by keeping them at a distance from the main leaf surface. If kept away from the cozy and nutritious surface, they may not thrive.

It would be foolish to look among these for a common primary function (a universal "adaptation") and then to consign the others to a lesser category. By "lesser" I mean that however useful they might be, these functions would be considered incidental relative to natural selection—what are sometimes called exaptations. A far more likely scenario is that different functions for hairiness drive selection under different circumstances.

Finally, a recent paper[12] suggests an odd hazard of waxy hairs that recalls the first topic of the chapter. Sunlit water drops have long been believed to pose a problem for leaves by focusing sunlight onto specific points and heating the points to lethal levels. They are thought to act like magnifying lenses that focus sunlight and ignite pieces of paper. Supposedly, that's a reason not to water your plants while they're in direct sunlight. Such water drops have even been blamed for causing burn spots on human skin. By making drops more nearly spherical, hydrophobic or superhydrophobic hairs should exacerbate the problem. But the danger to either leaves or skin may be minor. For a droplet of water to focus light on the surface on which it rests, the light must strike it at the low angle of 23° from the horizon, something that would happen in early morning, late afternoon, or at very high latitudes. Were a leaf to be angled so the midday sun struck it at that angle, the drop would have already rolled off. For nonhairy leaves, what appear to be burn spots more likely come from various kinds of bad chemistry rather than bad optics of water drops.

Hairs, though, could make this nonproblem into a real one, one intimately related to their hydrophobicity. Hairs can hold a drop far enough from the underlying surface so that light focuses on that surface when striking at angles near 90°—the midday sun for a horizontal leaf. At this point the significance of this tantalizing idea awaits illumination.

Even More Multifunctionality: Thickness versus Life Span

A functional attribution asserted in chapter 8 runs into the same complication when we try to work back from function to evolution, trying to deduce the purpose of something in an ultimately purposeless system. As a general rule, leaves in dry habitats are thicker than those from less dry habitats—increased thickness has long been considered a prime element of the suite of characters referred to as xeromorphic.[13] Thick leaves do stay less hot under realistic conditions of direct sunlight and very low and fluctuating wind. But thickness shows a strong correlation with the lifespan of leaves. Ones shed annually are almost invariably thin, which makes the same functional sense as our use of cheap material and shoddy construction in single-use throwaway products. So is thickness primarily a response to the danger of high temperatures, or primarily to the economics of low productivity and the consequent impracticality of an annual leaf crop? Or, to play devil's advocate, are both incidental to some third function? One suggestion is that extra volume relative to area accommodates substances, physical and chemical, that discourage herbivory. That would be of greatest value where replacement cost relative to productivity is high—such as where water is scarce.

WINTER. HOW CAN A LEAF endure its cold, snow, and ice? Not all trees, not even all hardwoods, evade the problem by shedding an annual crop of leaves: as fall turns to winter, those nondeciduous magnolias and hollies in my yard become ever more conspicuous.

Brief residence in the freezer section of an ordinary refrigerator does obvious damage to vegetables—as figure 10.1 shows. But the leaves of the daffodils in our yard endure air

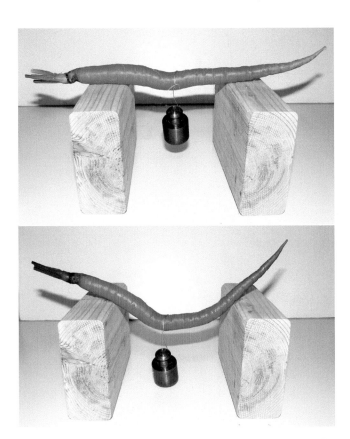

FIGURE 10.1. Slow freezing—in a home refrigerator—generates ice crystals large enough to disrupt the pressurized cells that give a carrot its crispness. Here, a carrot before freezing and after thawing. Celery responds even more dramatically, but the product is too ugly to illustrate. Commercial freezing must be done very rapidly (a process called flash freezing) to keep crystals small enough to have any chance of preserving texture. The blocks are 4 inches apart; the weight has a mass of 100 grams.

Staying Unfrozen

10

temperatures almost as low, with their flowers sometimes sticking up through snow. Yet the experience evidently leaves them unperturbed. New buds of trees and shrubs develop in anticipation of the spring, while at our local farmer's market fresh winter greens—collards, kale, mustard, cabbage, and so forth—appear each week.

Active organisms depend on water in its liquid phase. Sure, some have specialized inert stages that endure desiccation—seeds, spores, dried rotifers, brine shrimp eggs—but these serve only as devices for dispersal or for waiting out waterless episodes. Since their medium remains liquid— it's a rare pond in a rare chill that freezes down to the bottom—aquatic creatures enjoy a better world. But terrestrial life bangs up against the unpleasant reality of global temperatures averaging only around 10°C (18°F) above the point at which water goes solid, temperatures that all too often and in all too many places drop all too far below that average.

What makes the greatest trouble isn't the impossibility of swimming in ice, or the difficulty of grazing through snow, or any drought-induced thirst. It's ice formation within the organism itself. The problems caused by freezing go far beyond anything that might be handled just by awaiting warmer weather. Ice formation itself makes most of the mischief—internal ice formation bumps up against a number of unpleasant physical phenomena:

- Ice floats, as we know from pictures of icebergs and our ice-containing drinks.* Ice occupies about 10 percent more volume than the liquid water from which it came and, if internal, can disrupt the elaborate structure of living material.
- Left until it slowly dries up, a strong salt solution develops impressively large crystals. Crystal growth, whether of salt or ice, typically pushes other solids out of the way—not a good thing if it happens in living material. The slower the growth, the bigger the resulting crystals.
- Ice cubes (fig. 10.3) ordinarily contain bubbles of gas. Dropping the temperature of liquid water permits ever-more gas to go into solu-

*In a sufficiently alcoholic and thus low-density aqueous solution, ice will sink. But no, you can't ordinarily tell if the whiskey for your whiskey on the rocks has been watered down: the alcohol level at which the density of gasless ice equals that of the liquor is a little above 50 percent, or 100 proof.

tion—until the water freezes, when most of the gas fizzes out of solution. That fizzing out is potentially disruptive when it happens, and it's once again hazardous when melting frees the bubbles to move around and coalesce.

- When water with dissolved material in it freezes, the ice tends to exclude all but the water. So the dissolved stuff gets ever more concentrated in the liquid that remains. That has serious consequences—in particular for the osmotic situation—for what remains.

No question, ice is Bad News; in effect it's toxic water, our earth's most common toxic substance.

How to escape? We animals can pick among a wide range of schemes. If you're big, well insulated, and sheltered, you need only turn down the me-

Do it yourself (after reading the footnote and looking at figure 10.2): You can contrive a youth-puzzling demonstration by making up such a mix with either denatured ethanol from the paint store or rubbing alcohol (most often isopropyl rather than ethyl alcohol). The ice (cubes or crushed) will sink but then slowly rise as the water melted from it increases the density of the mix below it.

FIGURE 10.2. Pieces of ice melting into a mixture of about 65 percent alcohol and 35 percent water. The ice cubes initially sink, but then their melting puts more water into the mixture below them, making it more dense. So they follow the resulting density gradient upward until no ice remains. The cubes are light blue; the alcohol-water mix resulting from melting and convection is a darker blue.

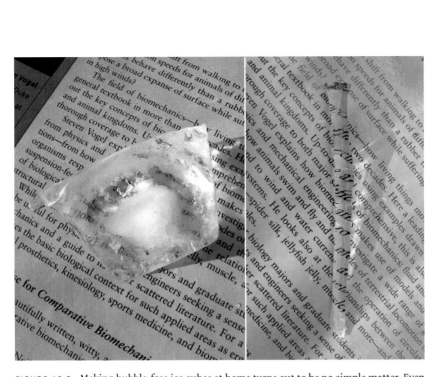

FIGURE 10.3. Making bubble-free ice cubes at home turns out to be no simple matter. Even with preboiled distilled water, some dissolved air remains and gets released. But notice, on the left, that that residual air makes bubbles in the middle—freezing inward from the edges chases gases into whatever liquid remains, until that last bit freezes. Icicles, by contrast, tend to be clear. Why? Their gradual, treelike growth means they freeze (as they grow) from the inside out, so dissolved air ends up in the general atmosphere rather than being trapped.

tabolism and quietly wait it out—as does a bear in its den. If small, cold-blooded, and semiaquatic, you go jump in the lake and again wait it out—as do many amphibians. No organism fills itself with salt-free water, so the lake will freeze before you do. That is, *if* it freezes, since in few places do lakes entirely freeze, insulated as they are by floating ice with its low thermal conductivity. If small and warm-blooded, you run your metabolic furnace just enough to keep body and soul together, meaning above the lethal freezing point—as do mammals that truly hibernate. And so forth.

Leaves can use none of these escape routes. Their basic surface-demanding function leaves them with lots of outside and little inside. That rules out metabolic thermogenesis, so their temperatures depend on their interactions with their surroundings. Surprisingly, some plants do generate sufficient heat to warm up appreciably by internal chemical reactions, much as we mammals and birds (and some flying insects and fast-swimming fish) do. But only unflat structures such as buds can do that, these last stealing a march (often in March) on the expanding and progressively shadowing foliage above them. The process costs stored energy, but it's apparently worth paying as the price for getting leaves out early enough to take advantage of full sunlight while it's available.

A few kinds of leaves supercool, most notably the leaves of bamboos—some don't freeze unless the temperature drops below about –20°C, or –4°F. The bamboos in my yard, planted before I bought the place (so I'm blameless—they're an invasive nuisance), keep their leaves through our winters, which once in a while get that cold. Some palms and a few other semitropical plants can do the same thing. Nonetheless, it's not a particularly common solution, limited largely to plants for which low temperatures are neither too common nor too extreme. Exactly what allows these leaves to supercool remains unclear.

Antifreeze

Nothing is easier. Add almost any solute to water, and its freezing point drops below 0°C (32°F). That's the main reason you can erode the ice on your steps by sprinkling with salt. A little melt at the surface of each salt crystal gives strong brine, and further ice dissolves into the brine—ice is soluble in water. Put a jar of jelly in the freezer, and the sugar in the jelly means you needn't worry about the jar cracking from expansion as ice forms. Or, as the Danes do with aqvavit, you can keep a bottle of hard liquor in the freezer without risk of solidification.

The relationship between the concentration of solutes in a solution and its freezing point needs some elaboration. Depression of the freezing point is one of a group of behaviors referred to as the colligative properties of solutions, mentioned briefly back in chapter 5. *Colligative* comes from the same root that gives us *collect*, *colloid*, *collagen* (glue-former), *colleague*, *college*, and *collage*, all referring to some sort of groupiness. Of these four properties, three have major biological consequences. Here, then, the collection of colligatives, the consequences of concocting solutions of solutes in solvent:

- Depression of the freezing point. That's the current event.
- Elevation of the boiling point. That, together with the first, is why you

add ethylene glycol to the water in your car's radiator. I've not heard that it matters to organisms otherwise.

- Reduction in vapor pressure. Adding solute makes the liquid in a solution less eager to go gaseous, so it evaporates more slowly.
- Change in osmotic pressure. Water moves from higher to lower concentration when it can, and solutes reduce its concentration (chapter 5, again). Thus they increase the pressure difference between a compartment of pure solvent (higher) and one with added solutes (lower).

The same rule applies to all: the change in any of the properties depends directly on the number of molecules of solute relative to the amount of solvent. A big molecule produces no greater effect than a small one.* Ethylene glycol (the main solute in your car's radiator), with molecules 5.5 times lighter than those of sucrose, gives, per unit weight, 5.5 times more freezing and boiling resistance than does table sugar. Put another way, an antifreeze consisting of lightweight molecules doesn't have to be so concentrated (on a weight basis) to drop the freezing point by the same amount. Nature makes wide use of a permutation of that trick. Take

---

*How can one count molecules? You have to weigh some amount of a substance and then convert weight to number. To do the latter, look up the so-called atomic weight of each atom in a molecule of the substance and add them up to get the molecular weight of the substance, assuming all the while that the weights are in grams. The colligative strength of a solution depends directly on the number of so-called gram-molecular weights dissolved in a given mass of solvent—one gram-molecular weight per kilogram is called a one molal solution, to simplify just a bit. (You aren't really counting atoms or molecules, which would give awkwardly high numbers, but a unit consisting of 6 times 10^{23} of them—Avogadro's number. That particular multiplier gives hydrogen, the smallest atom, a convenient gram-atomic weight of 1.0.)

How much does shifting from pure water to a one molal solution alter things? It depresses the freezing point by 1.86°C (3.35°F). It raises the boiling point by 0.51°C (0.92°F). It generates an osmotic pressure (against pure water) of 2.26 megapascals, or 22.4 atmospheres. At 25°C (77°F), it lowers the vapor pressure of an aqueous solution from 3168 to 3063 pascals (by Raoult's law). (Vapor pressure depends a lot on temperature, reaching 1 atmosphere for pure water at the boiling point at sea level.)

One other note: if the solute molecules break up when dissolved, then you have to count the pieces, not the original number. Thus sodium chloride, NaCl, ionizes completely in aqueous solution, forming Na+ and Cl–, so each molecule gives twice the normal bang per buck for any of these properties.

a big molecule, for instance starch, and break it up into its component small ones (for starch, glucose), and all four colligative effects become stronger. For instance, a plant cell could become more turgid by breaking starch and taking advantage of the greater water-sucking osmotic effect of the resulting sugar.

Depressing the freezing point by adding solutes looks like a small thing: full-strength seawater freezes at –1.85°C, or 28.67° F. Ordinary plant cells, those of young leaves, freeze at –0.56°C (31.0°F). These depressions, though, get amplified by the way that ice formation leaves a more concentrated solution behind, what's called fractional crystallization."* Moreover, in the presence of solutes, the frozen product doesn't come out hard and homogeneous but becomes a sludgy mix of ice and concentrated solution, which is ordinarily less damaging. As will become increasingly clear, freezing is a remarkably complicated business, decidedly more so than that other familiar phase change, vaporization.

Bottom line: just adding solutes works, but it faces several problems. To get a substantial freezing-point depression, a lot of solute has to be added. Thus adding salt, really Na+ and Cl– ions, may look good: a grain of salt (yes, that's a real if outmoded unit, not just an expression) gives as much depression as 1.6 times as much ethanol (once used in cars) or 3.2 times as much glycerol (used as antifreeze by many insects). But enzymes tend to be quite particular about ionic concentrations, so salts present especial difficulties. Furthermore, small organic molecules diffuse all too readily, especially through cell membranes. Methanol, almost as good as salt (1.1 times as much needed), is the most mobile of water-soluble organics. Sucrose stays put fairly well, but producing the same effect with it takes almost twelve times as much.

Worse, perhaps, freezing-point depression and osmotic pressure are both colligative properties and so have inseparable bases—going for one

---

*You can concentrate the alcohol in a solution by freezing out some of the water as well as by distilling and collecting the distillate. I heard that old-fashioned applejack was made from hard cider in this way. So I tried it once; the product was undrinkable even by us graduate students. In distillation one condenses the vapor, so bad-tasting large molecules are left behind and discarded; in fractional crystallization one gets rid of just the water, concentrating all the heavy stuff. So for the latter another step is needed, either treatment with some absorbent solid or else a distillation.

gets you stuck (collaged) with the other. And messing with osmotic pressure means other trouble, so you trade one disability for another.

> **Do it yourself:** Food freezers fail from time to time; where I live the electricity supply goes out all too often, sometimes with no obvious provocation. A simple device gives you a view of how a lot of organisms cope with freezing temperatures, and at the same time yields something that helps you endure powerless episodes. Water has a relatively high heat capacity (chapter 8) and, better, a high heat of fusion, so a small volume can absorb a lot of energy, and absorbing energy that leaks into the freezer is the name of the game. But the lovely heat of fusion comes into play only when the temperature has risen to 0°C (32°F), too high for most stored food. So storing pure water-ice in a freezer gives you only a buffer against loss of cooling of its ordinary heat capacity. Addition of solute, though, can lower its freezing point enough to enlist the heat of fusion. I've found that a mixture of alcohol and water does the trick nicely—a couple of half-gallon or gallon plastic milk containers hold the product. A 20 percent solution (by volume) of 95 percent ethanol forms an icy sludge at the normal freezer temperature of −18°C (0°F). It then gradually melts as the temperature rises, absorbing energy. (Starting with 70 percent isopropyl rubbing alcohol means you'll have to go up to 35 to 40 percent.) If you have a separate freezer, the stored jugs of this frozen mix can additionally serve as portable freezing elements for a failed refrigerator.

Incidentally, you take advantage of the final colligative property, boiling-temperature elevation, to measure the concentration of a sugar solution. That's the principle behind candy thermometers. As water boils off, the boiling temperature rises, so the thermometer can be marked with indicators for syrup, soft balls, hard balls, and so forth. In the same way, you determine the end point with a thermometer when making maple syrup by boiling most of the water out of the sap.

Keeping Ice Extracellular

Keeping ice outside the main machinery of an organism seems to be a general requirement. To declare it universal would overstate the case only a little. A couple of animals can apparently tolerate intracellular ice—a nematode worm and an insect, but these are the rare exceptions.[1] Plant cells have been shown to tolerate intracellular ice, but the requisite conditions probably don't occur outside the laboratory. Cooling must be rapid enough to keep the crystals of ice submicroscopic or to produce a glassy rather than a truly frozen state. That rapidity demands heroics such as plunging thin samples into liquid nitrogen, below –196°C (–321°F) (or liquid air, below –183°C or –297°F, much the same thing).[2] Under normal circumstances, freezing kills plant cells, too.

Ice crystals disrupt the organization of cells. Until the advent of electron microscopes put an end to what had already become an anachronism, introductory textbooks described cells as bags of protoplasm containing various suspended organelles. By the mid-'50s, cells had filled up with all manner of particles and membranes, leaving almost no space for protoplasm, which was retired to well-deserved oblivion.[3] Only biochemists equipped with homogenizers and test tubes could still turn their backs on the high degree of internal cellular organization.

How can ice formation be kept extracellular? That's not quite as hard as you might guess. Cool water gently, and it may not freeze at 0°C (32°F)—it commonly supercools down to some temperature between 0° and –39°C (+32° to –38°F). Freezing then occurs almost instantaneously, initiated by almost any particle or perturbation. In the laboratory part of a physical chemistry course, I learned that what we call a freezing point is really determined as a thawing point, a more reliable and repeatable datum. Freezing needs an initiator, a nucleator, it's called, and if the initiator is extracellular, then the ice forms extracellularly.

When it does happen, ice formation has an upside. Just as thawing requires a source of energy, forming ice releases energy, quite a lot of energy.* Making ice requires that the original water lose as much energy as if it cooled by no less than 80°C (140°F). So ice formation slows cooling considerably, enough so leaves freeze further more slowly than otherwise—up to a few hours in extreme cases. If the stimulus for freezing

---

*This is the so-called heat of fusion. For water at 0°C (32°F), the heat of fusion is 334 kilojoules per kilogram. While far from trivial, that's much lower than the heat of vaporization—2206 kilojoules per kilogram at 100°C (212°F), nearly seven times more—mentioned in chapter 8.

doesn't persist, as might be the case when air movement picks up after a while during a clear, cold night, little ice may form. Farmers sometimes spray crops and fruit trees with water before and during cold nights—a wet or wetter surface water means slower icing.

Initial ice formation during a sudden exposure of a leaf to a subfreezing temperature can even cause its temperature to briefly rise above that of its surroundings. Warming from the energy released can limit further ice formation. Around dawn, when both air temperature and air movement hit their nighttime minima, that warmth can significantly reduce freezing.

At least one alpine plant, on Mount Kenya, depends on this thermal buffering to limit the extent to which it freezes. Its flowering structure contains a lot of water, which heats up during the day. Sometime during the night, the water reaches the freezing point and begins to freeze. But the release of heat in the process keeps its temperature near 0°C (32°F), even though the air is substantially colder. Temporary amelioration, yes, but day does follow night with dependable regularity.[4]

Why don't growing ice crystals extend into cells and eventually expand intracellularly? Cell membranes happen to provide excellent barriers to crystal growth. Recall the instability of very small air bubbles, the way they prefer to self-destruct as the extra pressure from surface tension urges the air to dissolve. Tiny ice crystals suffer an analogous instability, although its physical basis differs slightly.* Crystals are reluctant to form without some nucleating element—again, why supercooling occurs, and, again, setting an effective minimum size. For ice crystals the minimum size of an "ice embryo" is about 20 nanometers. (A nanometer is a millionth of a millimeter; 20 nanometers is within the size range of macromolecules.) That may be smaller than the water-filled spaces within cell

---

*One is looking at quite a general phenomenon, though. Making surface ordinarily takes energy; put another way, surface prefers not to exist. Surface tension, recall, has dimensions of force per distance. That's equivalent to work (or energy) per unit area of surface, the energy needed to make surface or that surface reduction liberates. The phenomenon applies to solids as well. A crack will not propagate until deep enough (critical crack length, about which more in chapter 13) so that the relief it affords some stressed object yields enough energy to keep it going. Break rods of various materials such as glass, fiberglass, or dry or fresh wood—the more jagged the resulting broken surface, the harder you probably had to work to make the break.

walls, but more important, it's larger than the pores that perforate cell membranes. Cell membranes thus provide barriers that even minimal crystals can't penetrate.[5] (But the outer surface of leaves presents less of a barrier. A bit of frost on the outside facilitates freezing on the inside.)

Furthermore, the growth of ice crystals within aqueous solutions has a paradoxically self-defeating characteristic. The crystals grow by addition of water molecules and the exclusion of almost everything else—other kinds of molecules don't usually fit into their structural lattices. As a result (and as mentioned earlier), solutes in the remaining solution get ever more concentrated. And the more concentrated the solution, the lower the temperature at which ice begins to form.

One more piece. Cell membranes may stop ice crystal penetration, but water passes through them with relative ease. That means adjacent extracellular ice formation can increase the concentration of dissolved material within cells. So if either intracellular ice crystal growth can be postponed or extracellular formation can be triggered, crystallization will continue to be extracellular.

The basic issue can be put in practical terms. How can sites of nucleation be controlled? Among plants, one common trick consists of expelling a considerable fraction of the water within cells before freezing has started. That leaves the cells with more concentrated fluid. That fluid is even more concentrated than we might guess, since then a larger fraction of the remaining water is left bound to subcellular structures. In a real sense, it's no longer a freezable liquid. And exporting water from cells dilutes the extracellular fluid, making it more likely to freeze. The process anticipates actual freezing. Of course, those dehydrated cells have less volume, so where stiffness depends on volume, where cells behave like our inflatable structures, they go limp. The frozen-food industry, no surprise, tries to freeze things as fast as is practical. "Flash freezing" is not just an advertising slogan but a way to leave little time for water export. It's the invention, many years ago, of Clarence Birdseye—yes, that's the origin of the brand name.[6]

The nature of the freezing process itself offers another possibility that ought to be mentioned. When water (unlike most liquids) freezes, it increases in volume. The crystalline structure of ice occupies more space than the loose and less ordered intermolecular bonding of water in the liquid state. As a general rule that we've met in various guises, on their own, physical and chemical processes don't ordinarily go in directions that require inputs of energy. Pushing aside other material takes energy, so basic thermodynamics tells us that positive pressure, any inward squeeze, will oppose any volume increase. Thus it will oppose the freezing of water. Similarly, boiling involves expansion—huge expansion—so the more pressure applied, the higher the temperature needed to get the pot boiling. Cooling systems in modern automobiles operate at pressures above that of the local atmosphere, so the liquid coolant can run hotter without breaking out into a boil. The result is a small increase in overall engine efficiency. (Do not—I repeat, do not—remove the pressure cap when an engine has been warmed up—a scalding mix of liquid and newly formed vapor may erupt with great violence.)

Within plant cells, unlike the situation within animal cells, pressures are ordinarily well above that of the extracellular (and atmospheric) world. The freezing point of the intracellular fluid of young leaves, mentioned earlier, is about –0.56°C (31°F). That can be converted to an effective concentration of solute—an osmolality. If the extracellular fluid contains no dissolved material, we calculate an osmotic pressure difference across the cell walls of almost 7 atmospheres.* So that's the pressure ice formation must work against, and it drops the freezing temperature by an additional tenth of a degree (C). Not much and of uncertain and unlikely significance, but it's worth noting in an account that explores physical possibilities.† Besides, once alerted to the possibility, someone somewhere might identify an instance of its use.

---

*We just use the data given earlier for freezing-point depression and osmolarity. The effective concentration is –0.56 over –1.86, or 0.3 molal. The osmotic pressure is then 0.3 times 22.4, or 6.7 atmospheres.

† This effect often takes credit for making ice skating practical. A skate presses down on the ice with great pressure, since the small area of contact concentrates the skater's weight. That might give localized, brief melting, providing lubrication that eases movement of the skate's runner. But the effect may not deserve such credit. If it were responsible, then skating would work far better just below 0°C (32°F) than at lower temperatures, which isn't what happens. This case against liquid lubrication is made by Joe Wolfe (2010) at www.animations.physics.unsw.edu.au/jw/freezing-point-depression-boiling-point-elevation.htm; he also does the full calculation using the Clausius-Clapyron equation from which (adjusting input pressure) I obtained that datum of a little less than a tenth of a degree.

Vitrification

As an alternative to exporting water, the cells of a leaf might keep it in some state in which it won't freeze. We usually think of water as coming in three phases, gas, liquid, and solid, but in the presence of other interacting substances, that's not the whole story. And cells are chock-full of other substances as well as submicroscopic structures. Faced with the risk of freezing, a cell can accumulate water-binding substances, in particular water-soluble carbohydrates. We do that when thickening a pie with cornstarch or stabilizing jelly with fruit pectin, both carbohydrates.

In the past two decades, ice control through a fancy implementation of this last approach has been recognized. It has long been known that gradually acclimatizing a plant—days and weeks, not just minutes and hours—often leads to a greatly increased tolerance of cold. Such long periods provide time for the mix of proteins synthesized by cells to change. In particular, a plant can turn on genes that encode so-called antifreeze proteins, a designation that doesn't do full justice to their roles.

Antifreeze proteins, like other proteins, can't play the colligative game—protein molecules aren't just large, they're so huge that persuading a large number of them to dissolve in a small volume of water verges on the impossible. But proteins range from highly hydrophilic ones that swell by absorbing water to hydrophobic forms that refuse ordinary attempts to coax them into solution. They interact in complex ways with water, as when a solution of gelatin, a hydrophilic one, forms a peculiar soft solid, a gel. We sometimes thicken foods with gelatin or eat it, flavored and sugared, as Jell-O. Another such protein-water gel forms when the liquid around a cooked fish is chilled. Prepackaged edibles mainly use algal proteins such as agar, but the gelling action remains the same. In effect, benign gelling discourages lethal freezing.

Removing water from aqueous solutions can have various outcomes. Sometimes what remains is a pure, nonaqueous liquid, while in other cases some solid, crystalline material accumulates. Antifreezes, both small-molecule and protein, may lead to a third possibility, one not so often recognized. Removing water and dropping the temperature can produce an amorphous—meaning noncrystalline—yet solidified product. We call the process vitrification, and the vitreous product a "glass"—

vitrum being Latin for "glass." The product may be soft at first, but removing ever-more water makes it ever more viscous until for all practical purposes it becomes a solid, even if it's not a properly crystalline one.

The way extremely rapid cooling could produce a glassy state in the cells of leaves was noted a few pages back, and that's what we've returned to. This glassy state behaves much like a fourth phase of matter. The transition from supercooled to vitreous lacks the characteristically sharp and specific temperature of freezing or boiling. In particular, the transition temperature varies with the speed of cooling. Vitrified solutions are not merely supersupercooled ("hypercooled"?)—their physical properties differ from both supercooled liquid and crystalline ice. In particular, they have their own heat capacities (the amount of heat it takes to change the temperature by a given amount) and thermal conductivities (the rate at which a temperature difference makes heat flow through a barrier).* Supercooled solutions range from fairly stable to ones so unstable that introduction of a crystal of solute produces an explosive wave of freezing. By contrast, vitrified solutions are as stolid as they are solid.

Sugar solutions can both supercool and vitrify. For simple sugars, those with a single six-carbon structure such as glucose, galactose, and fructose, both the supercooled and the vitreous states can be stable for long periods of time. Corn syrup, mainly glucose, sits in its bottle on your shelf quite unconcerned that it's a supersaturated solution. I encountered this oddity once, when I needed to know the viscosity of the corn syrup solution in a small flow tank. It seemed no great problem—I measured the rate of fall of a small, steel bearing ball and from that figured the viscosity with the simple and tried-and-true Stokes' law. I then looked up the concentration of glucose at saturation in a handbook and compared that with the concentration of sugar I calculated from the nutrition information on the corn syrup bottle. Total mismatches—both my measured viscosity and result of the nutritional calculation were a hundred times greater than a saturated solution should have given. Both meant that even though no crystals formed, the concentration of sugar was well above saturation. The puzzle finally yielded with the aid of *The Fannie Farmer Cookbook*. To

---

*But the vitreous humor that fills our eyeballs happens to be a gel rather than a glass; the name recognizes its transparency rather than its physical state.

produce sticky cake frosting, it calls for a simple six-carbon sugar; to produce brittle frosting it specifies a double, twelve-carbon sugar, such as table sugar, ordinary sucrose. Commercial corn syrup, a simple sugar, sits there as a liquid, quite unconcerned that it's vitreous or supersaturated.

I should have guessed. On several earlier occasions I had participated in making a peculiar New England confection called sugar-on-snow. The process consists of boiling maple syrup considerably further than usual, pouring small dribbles on a local snowbank to chill and harden, and, sitting on the snowbank, eating the sticky result with a fork. Aficionados offset its cloying sweetness with intermittent nibbles of cucumber pickles, and they offset both (and the cold) with lots of hot coffee. The product has no commercial version—maple syrup contains too much sucrose, so when left to its own devices the vitreous form slowly transforms itself into stable, crystalline, maple sugar candy.

Vitrification happens within the cells of a large variety of both animals and plants, ones that tolerate seasonal exposure to extreme cold. Antifreeze proteins, similar in both animals and plants, promote vitrification, as do a variety of chemicals, most of them highly hydrophilic. Both the double sugar trehalose and the triple alcohol glycerol (aka glycerin or glycerine) have repeatedly been pressed into service. They may rise to remarkably high concentrations as creatures gradually acclimate—doing what's called cold hardening by botanists and horticulturists.

The most extreme cases—think of insects in the tundra or the leaves of evergreens at the northern tree line—reduce the volumes of their extracellular as well as intracellular fluids. What's left after that dehydration may then vitrify. Investigating the vitrification of biological material has drawn considerable effort in the past few decades, and we now know much about how to make it happen. The impetus comes from a completely practical matter: preserving living material for transport or storage, whether sperm, ova, entire organs, or even whole embryos. In two words, vitrification works.[7] What we know less about—not that the subject passes unnoticed—is vitrification as it happens under *natural* circumstances. It's nothing if not a potent process. According to a recent report, a beetle that finds northern Alaska tolerable can supercool down to –58°C and can survive by vitrification at still lower temperatures, down to –100°C (–150°F) in the field and as low as –150°C (–240°F) in the lab.[8]

The Bottommost Line

Freezing, or avoiding freezing, stirs up more complications than one might have thought—more, certainly, than I guessed before I began writing this chapter. The obvious device capitalizes on one particular colligative property of solutions: reduction in freezing point as solute is added, obvious at least to anyone who ever studied physical chemistry. But it plays a limited role in a leaf, most likely because of life's intolerance of changes in another colligative property, osmotic pressure. Furthermore, the inseparable link between osmotic pressure and freezing-point depression means that tinkering with such things as the sizes of solute molecules provides little or no help. Applying pressure raises boiling points very nicely, and plant cells commonly have high internal pressures. But pressure changes do precious little to freezing points.

What can be and is done by the leaf to survive the winters of nontropical land depends on some combination of about four (at least four) devices: limiting by dehydration the water that might freeze; ensuring that ice forms extracellularly; controlling the size of ice crystals with appropriate additives; and promoting supercooling and vitrification with these and other additives.

All this talk about ice brings up a major contemporary conundrum: the climate change business that all of us are all too tired of hearing about. We might expect that a shift in temperature of a couple of degrees won't really make much difference. If concentrated in the extreme north and south, warming might even improve the habitability of the earth for humans. One fly in the ointment, though, emerges from a peculiar combination of factors, ones that have come up in this chapter. The earth's average temperature, about 10°C (50°F), is close to the freezing point of water. Liquid and solid water have different properties. And the freezing process is nothing if not peculiar.

As a result of that low average temperature, at any given time a great deal of the earth's water is frozen. It reflects radiation outward from the earth, limiting solar heating—so less ice means more heating which means less ice still, at least until some compensating factor emerges. I find that positive feedback (recall chapter 5) a truly scary prospect. Melted ice is less dense than seawater, so the melt can flow out over bodies of water, with odd effects on ocean currents, which play a critical role in determining the climates of the different parts of the planet. Less often mentioned, ice excludes solutes as it forms. Icing of seawater thus generates a brine just underneath the ice that's denser than the seawater farther down. The brine's sinking produces oceanic currents, so suppressing ice formation turns that off, with further uncertain but probably synergistic consequences. With the earth's average temperature just above the freezing point, even a small rise in global temperature will melt a lot of its ice. We've begun to worry about the consequent rise in sea level. But in part that may be because such a rise is the most predictable consequence of the warming. Others, as important or more so, are less easy to anticipate.

Finally, recall that little air dissolves in ice. So well-aerated water disgorges about 2 percent of its volume in the form of vapor bubbles when it freezes—as in the ice cube of figure 10.3. Does that matter? Working on arctic plants, Scholander and his coworkers drew attention to the phenomenon over half a century ago.[9] Its importance, if any, remains uncertain. People concerned with cryopreservation know that solutions should be degassed, but freezing-induced outgassing in nature hasn't received much attention. Perhaps it's unimportant. Air bubbles should remain extracellular—if they're small enough to penetrate the pores in cell walls and membranes, they're too small to persist. Still, wherever they form, bubbles will make ice a little less dense than it would otherwise be—perhaps 10 percent less rather than 9 percent less than the density of liquid water at the freezing point—which might well have significant effects on such things as buoyancy. Once again, physical possibilities need to be raised, even if in the end we do no more than definitively dismiss their relevance.

OF WHAT VALUE is all that photosynthetic machinery, all those schemes for moving gases and water, if a tree can't hold its branches outstretched, if its leaves can't extend from their twigs and intercept light? Our paradigmatic leaf has to expose its surface to the sky, not collapse and sag downward like a flag in a lull. So we turn, where we might have started, to these mundane mechanical matters, ones more familiar to structural engineers than to biologists or physicists.

Support—that's the subject; material—that's what's available; structure—that's what must do the job. For us, bones, tendons, and muscles take on the structural tasks. Proteins provide the basic materials, either alone in tendons and muscles or in composites with calcium salts to produce bones of adequate stiffness. For plants, by contrast, stiff cell walls, water under compression, and wood provide structure. Carbohydrates and liquid water are the basic materials, quite different stuff used in quite different ways. So yes, humans can survive on a diet of plants—indeed, most humans alive at present do so either by choice or by necessity. But strict vegetarians tickle the edges of protein deficiency in a way that flesh eaters don't.

Material

One carbohydrate carries the load in an especially literal sense: cellulose.* And, in the same literal sense, it carries our own loads. Cotton (from the cotton plant, *Gossypium*), linen (from flax, *Linum*), jute (or burlap, from *Chorcorus*), sisal (from *Agave*), coir (from a coconut, *Cocos*), and some others—most of the natural fibers we've long used for cloth and cordage share two features. Except for wool and silk, all come from plants, and are all composed of the same material, cellulose. The first artificial fibers, cellulose nitrate (later

---

*I'm using *cellulose* as the generic term for most of the fibrous material of most plants, lumping cellulose in the strict sense with various hemicelluloses, pectins, and so forth.

replaced by the nonexplosive cellulose acetate) and rayon, came from that same source. As a glance at any tree confirms, cellulose is a great structural material. Curiously, it consists of nothing more than a bunch of simple sugar units, the familiar glucose molecules that photosynthesis generates, chemically bonded to one another in chains. Starch looks almost the same when its structure is drawn on paper, but the bonds between its sugars differ just enough so that starch can be dissolved in water, while cellulose resists dissolution. It resists so well that trees laugh off rainstorms, we build long-lasting houses of harvested wood, and natural recycling relies on insects and microorganisms.

Cellulose itself isn't the basic structural material of plants quite the way steel is of a bridge or brick of a wall. Whatever its intrinsic virtues, it always forms part of a composite material: fibers of cellulose glued together in a matrix of a softer material, lignin. That's how nature most often arranges her structural materials. We may make things of pure metals, pure at least at a microscopic level, even if alloys at the molecular level, like bronze (copper and tin), steel (iron, carbon, and various other things), pewter, or brass. Our plastics are also mostly pure materials. But nature uses composites—for materials pushed against, always; for materials pulled on, usually. We've long woven materials from fibers: baskets and fishnets perhaps first, wearable fabrics later. But nature weaves nothing in the in-and-out, over-and-under way that we make a material capable of taking loads in two dimensions from fibers that can be loaded only lengthwise. Instead, she keeps her fibers together, randomly or in layers, with glue. The cell walls of leaves, then, consist of cellulose fibers in a feltwork held together by lignin.

(Not that we don't sometimes play the same game, and we're now playing it more often and more extensively than ever before. Fiberglass— strands of glass in an epoxy or other glue—has long been familiar. So we're all familiar with how differently it behaves from an equivalent sheet or rod of glass—imagine a vaulting pole made of glass instead of fiberglass. The newest of our big airplanes use composites for most of their basic shells.)

Cellulosic stuff, now looking at the whole composite, withstands greater loads when pulled on—when loaded in tension—than when pushed against—when loaded in compression. Both trees and their leaves, like all but the smallest terrestrial organisms, face the full force of

> **Do it yourself:** If you make a paste of wheat bran as fiber and egg white as glue and then bake the product, you'll have made a proper composite. It happens to have the additional virtue of edibility, so your teeth can assay its mechanical properties. While at it, you might sculpt significant shapes, flavor the material, and so forth. Baking times (try 300°F) will vary depending on the thickness of what you make and how much moisture you want to retain.

gravity without the buoyant offset of aquatic life. In gravitationally challenged structures, compressive loading must be as great as tensile loading. So keeping anything intentionally in tension becomes an interesting trick. If the cellulose-lignin composite takes the tension, what takes the compensating compression? Put another way, what does the task we assign to our bones?

Supporting Leaf Cells

For leaves, the compression-resisting material consists of nothing more than liquid water. That's pretty strange to us creatures accustomed to a technology centered on concrete, metals, and slices of dry wood. As mentioned quite a few chapters back, both gases and liquids (as well as solids) withstand compression—even if gases don't resist tension. But they take compression differently. Gases get notably squeezed in the process—double the compressive (positive) pressure, and gas volume then halves. Liquids, water as much as any of them, change volume very little, even under monumental loads.* Even if water were considerably more compressible, it would be adequate for what plant cells ask of it. What's critical is boxing up the water properly. That, though, asks nothing heroic either from the cellulose-lignin composite material of their cell walls or from the mix of substances in their semiliquid interior. As a supportive

---

*Double the pressure on water from 1 atmosphere to 2, and it responds by decreasing its volume by a mere one part in twenty thousand. The relevant datum is its bulk modulus, the ratio of pressure change (Δp) to volume change relative to original volume ($\Delta V/V_0$).

element, the basic structure of a plant cell resembles a water-filled balloon—more precisely, a balloon with a stretch-resistant fabric wrapped around it, as in figure 11.1. Take away the water, and what do you have? In

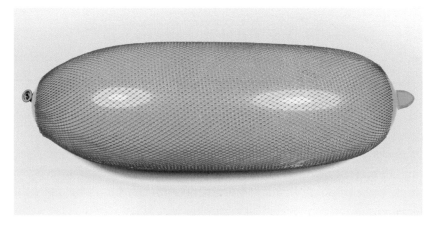

FIGURE 11.1. A cylindrical balloon has been inflated inside a tube with helical fiber reinforcement—one that came with some cherry tomatoes from the supermarket. This helical crossed-fiber arrangement is the one most often used for pressurized cylinders in living systems, whether plant cells or shark bodies.

essence, you have a cork, fine for stuffing down the necks of wine bottles, but not for building tall trees that stand up to storms.

We call these water-filled balloons, widespread among aquatic animals as well, hydroskeletons. Most of the animal examples are structures much larger than plant cells. Sea anemones; worms; the tiny tube-feet of starfish, sea urchins, and the like; crabs and lobsters just after they molt; sharks when they're swimming—all depend on water compressed to higher-than-outside pressure within some outer wrapping. The most familiar one plays an immediate role in our own reproductive success. As you've probably guessed, it's the erectable penis of male humans (and of many other but not all mammals, as well as of lots of other vertebrates). Incidentally, the internal pressure in an erect penis (the measurements were done on dogs) far exceeds blood pressure. Blood pressure does the initial filling, but some sphincter muscles then go to work, isolating the organ and running the pressure up to about a million pascals, 10 atmospheres. That's close to a hundred times our resting blood pressures—extraordinary for animals if merely ho-hum for plants.

Boosting the pressure in a plant cell takes no elaborate machinery such as a heart or a sphincter. A little more dissolved material inside than outside, and osmotic pressure does the job—recall that limp carrot in the last chapter, its cellular hydroskeletal balloons punctured by ice crystals. Even a modest concentration of solute produces lots of pressure: seawater inside and fresh water outside would generate over 20 atmospheres, 2 megapascals. The liquid in a typical plant cell, as the previous chapter noted, freezes (thaws, really) at about –0.56°C (31°F). That corresponds to an osmotic pressure—colligative properties being inextricably linked—to nearly 7 atmospheres, assuming pure water outside, which sap very nearly is. If your recipe calls for wilted lettuce, just briefly soak a few leaves in a salt solution. If you want your grated zucchini to be less watery, salt lightly, toss, and press out the water in a colander after half an hour.

Seven atmospheres sounds like a big pressure difference for a cell wall to withstand, particularly for walls less than a hundredth of a millimeter thick. No, we needn't invoke some spectacular tensile strength—a caution for aspiring biomimeticists. We simply return to the geometric argument made in chapter 7 when looking at how surface tension can withstand huge pressure. If the surface, the air-water interface, is very sharply curved, its tension resists very high pressures. That sharp curvature can happen if pores are very small, in effect when the system is tiny. Recall that before moving to surface tension, we applied the rule, Laplace's law, to automobile tires and blood vessels. Now we're back to that kind of system, one with a stretchy solid rather than surface tension as the tension resistor. Instead of figuring the maximum pressure across an interface from surface tension, we can calculate the tensile stress on the cellulose-lignin composite from the pressure difference.*

*Recall that for a sphere, which curves in two directions,

$$\Delta p = \frac{2T}{r} \text{ or } T = \frac{\Delta p r}{2}.$$

Instead of looking for tension, with dimensions of force per length, we're after stress, with dimensions of force per area. That's the relevant variable for a real material rather than an interface, which has no thickness. So the latter equation becomes

$$\sigma = \frac{\Delta p r}{2 \Delta r}.$$

σ is tensile stress; r, as before, is the radius of the sphere, or (strictly) the radius of curvature of its wall; Δr, the new variable, is the thickness of the wall.

For a cell with a radius of, say, 50 micrometers (two thousandths of an inch), a wall that's 5 micrometers thick, and a pressure difference of 700,000 pascals (7 atmospheres), the tensile stress comes to 3.5 megapascals. "Mega" makes it sound huge, but it's really not. A material that's stretched breaks at a maximum tensile stress, referred to as its tensile strength. Pulling dry wood (pine) in the same direction as its grain, the tensile strength is around 200 megapascals, while for pure cellulose it's 750 megapascals. Both figures are vastly above the 3.5 megapascals of osmotically generated stress.[1] So we need not lose any sleep worrying about exploding leaf cells—they've got strength either to spare or to invest in less idealized, nonspherical geometries.

Supporting Leaf Blades: Compression and Tension

Hydroskeletons rarely achieve great rigidity. That may explain their relative scarcity among large, gravitationally challenged terrestrial systems—at least relative to their widespread use by both tiny and fully aquatic creatures. Only very small leaves depend entirely on them to hold their shape. All the familiar ones gain support from their veins and their leaf stems, or petioles. Of course, veins act as conduits as well as supports, and figuring out which features serve which purpose can be tricky, especially since nothing rules out dual-function elements.

So how might a vein or petiole provide support? Here we enter a realm of engineering and things so ubiquitous that we rarely ask how they work or what the rules are that underlie their designs. I'm talking about beams, all those solid or weblike structures we see under floors or roofs. Most of our wooden ones are rectangular in cross section, narrow crosswise and wide ("deep") top to bottom—so-called two-by-eights, two-by-tens, and so forth. The numbers stand in for distances that may once have been actual inches; progress hasn't just cheapened currencies.

One particular genus of beam is of particular interest here; it's called a cantilever beam, a word of ancient and uncertain origin. Most often the term describes a beam that protrudes more or less horizontally from an attachment at one end, as in figure 11.2a. It may be loaded at its far end, loaded along its length, bearing only itself as the load (self-loaded beam), or it may resist some combination of loads. The term *cantilever*

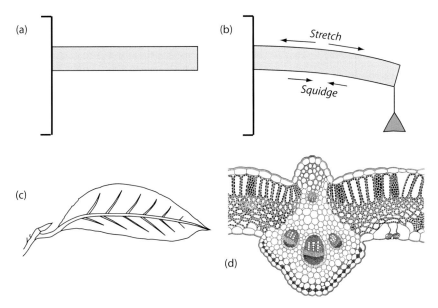

FIGURE 11.2. A flexible cantilever beam, one that protrudes outward from a support, unloaded (*a*) and loaded (*b*). Mechanically, most leaves (*c*) form cantilever beams of just this ordinary sort. A cross section (*d*), here of the midrib, an extension of the petiole, has large, thin-walled cells at the bottom that can resist compression like inflated tires. Near the top, the thicker walls of smaller cells have lots of lengthwise cellulose fibers, so they can act like ropes and resist tension—poking at a petiole or midrib with a pin will expose their orientation. The cross section comes from von Sachs's classic book, the one that contributed to several previous figures.

distinguishes it from other genera of beams such as the simply supported beam, which runs between supports at its ends. No mystery about the importance of cantilevers to the present story—most leaves protrude from stem, twig, or ground in exactly this way. Less obviously (and as we'll return to later), a whole tree that protrudes from the ground and feels a horizontal wind also acts as a cantilever beam. As anyone who does carpentry will immediately appreciate, making good cantilever beams demands more skill and care than making simply supported beams.

Consider the way the beam in the figure bears the load at its far end. Since nothing in the real world achieves infinite rigidity, the load bends the beam—at least a little. That has a peculiar consequence. The load stretches the top of the beam and compresses ("squidges" sounds better to me) the bottom, as in figure 11.2*b*. So the top of the beam finds itself loaded in tension, the bottom in compression. Most ordinary materials respond in quite different ways to the two loading regimes.

Ropes and cables can withstand only tensile loading. For that reason they're much easier to analyze than are beams. What's relevant here is that a rope, cable, lengthwise fiber, or any such tension-resisting specialist serves splendidly for the top of a cantilever beam—however hopeless it might be as a bottom. By contrast, bricks, cement blocks, or substantial chunks of metal (all heavy and thus costly in terms of self-loading) work properly for compressively loaded bottoms. Most often, though, we hide the distinction between loading regimes by making the whole structure of wood or steel, materials that take either kind of load without obvious disability.

Back to leaves. All this talk about cantilevers says something about how petioles and major leaf veins should be designed, about the factors that determine the structural arrangement in figures 11.2c and d. They ought to have fibers (of some cellulosic sort) running lengthwise near their tops and some more compression-resisting materials on their bottoms. What might do for the latter? If great rigidity were needed, something really hard—like nutshells, for instance—would be the obvious choice. But the supportive systems of leaves don't provide much more than the rigidity minimally needed to hold them out under their own weight. A small amount of snow or ice pushes a leaf far downward from its normally skyward orientation. As already mentioned, hydroskeletons do fairly well as supportive systems if the requirements for rigidity aren't too great. No surprise, then—the lower parts of petioles have few lengthwise fibers and a lot more of the water-bomb-like cells we described in leaves. Their walls may be somewhat thicker—rigidity still counts for something—but resisting most of the compression clearly falls to water rather than to any biosynthesized material. At the least, it's cheap, even free.

Supporting Leaf Blades: The Shape of the Beam

Old buildings were constructed with round or crudely squared beams to support floors and roofs. That's probably not a sign that our forebears misunderstood structural engineering but rather because trees are round in cross section, so hewing beams by hand was a lot of work—the same reason why log cabins were made of minimally trimmed logs. Nowadays our beams are deeper than they are wide or else I-beams of metal, similarly asymmetrical. Wide beams we reserve for flooring, roofing, and shelving, where function depends directly on width.

But how wide, how deep, how long? Engineering handbooks (and websites) provide formulas for the bending resistance of beams of every practical cross-sectional geometry and under practically every loading regime.[2] They start with the stiffness of a beam's material rather than its strength, because we care more about effective rigidity than about ultimate failure. A convenient measure of the resistance of a beam to bending is the sag (really lack of sag) at the end of the beam. That sag varies in ways prescient with practical significance,

- with the cube of the length of the beam sticking out from the support—twice as long gives eight times the sag;
- inversely with the cube of the depth of the beam, top to bottom—twice as deep gives an eighth as much sag;
- inversely with the width of the beam, side to side, neither squared nor cubed—twice the width means half as much sag; and
- inversely with the stiffness of the material of which the beam is made—twice as stiff means half the sag.*

The match of this idealized analysis and reality isn't all that tidy. Figure 11.3 shows several real, live petioles; for the most part they don't turn out to be vertically elongated ellipses or rectangles. More common are sections that are nearly circular, often with lengthwise grooves on their tops. Yes, deep ones do exist, most conspicuously in the poplar family: white

---

*For example, for a cantilever beam that's elliptical in cross section and loaded uniformly, that, is with the same load on each element of its length, the formula is

$$y = \frac{8Fl^3}{\pi E d^3 w},$$

where y is the downward sag of the end, F the load, l the length of the beam, and d the depth and w the width of the beam. E is the stiffness of its material, also called Young's modulus of elasticity. If the beam's weight is negligible and the load is concentrated at the end, the sag is several times greater, or

$$y = \frac{6.8Fl^3}{E d^3 w}.$$

Both formulas tacitly assume quite a lot: for instance, uniform composition and uniform response to tension and compression, assumptions that work far better for steel beams than for the veins and petioles of leaves.

Do it yourself: If you're feeling sufficiently compulsive, you can verify these proportionalities by applying weights to beams of varying lengths and cross sections. Cheap cylinders and other shapes of adequately uniform properties can be found in the pasta section of the nearest supermarket, although their fragility and small sizes may demand some manual delicacy. Simply supported beams (supports at the ends) give less trouble than cantilevers and behave in basically the same way. Bent lengths of thin wire (unravel some electrical cord) make decent weights—weight varies directly with length. The few strands of pasta diverted to experimentation shouldn't be missed when the rest serve their intended purpose. Or eat them as well.

FIGURE 11.3. Four petioles, all acting as cantilevers. Many, such as the dogwood, sweetgum, and elephant's-ear here, have grooves on their upper surface. These allow the petioles to twist fairly easily without compromising their resistance to bending downward. White poplars (and their near relatives) have petioles that are flattened side to side, allowing them not just to twist but also to bend side to side while resisting up-and-down bending.

poplar, cottonwoods, and aspens, as in the last panel of the figure. Our analysis seems to have missed something nontrivial. Part of the gap in the picture must be the presumption in classic beam theory that the material of a beam is uniform throughout it. Veins and petioles don't come close to meeting that condition, the point of the preceding section about local specializations to resist compression or tension. But that's not the whole story, and we'll return to the problem in the next chapter, resolving it by considering problems trickier than anything gravity imposes.

Treating leaves as beams helps us understand the veins of their blades as well as their petioles. Depth—that's what confers stiffness in the face of forces that would bend a leaf's blade downward. Veins might protrude above the blade, be centered on the blade's surface, or extend below the blade with nearly the same mechanical effectiveness. Extending below is most common in leaves that have thin blades, perhaps because the thin membrane of the rest of the surface can assist well if loaded in tension, as it would be on top, but only poorly if loaded in compression, as it would be on the bottom. Or perhaps veins protruding above might slightly compromise photosynthesis. Or such veins might encourage accumulation of water droplets, spores, dirt, and such on the top, material that would simply fall from the bottom.

Virtual Beams

A little fancier than beams but working much the same way are trusses. They're beams made of multiple structural elements, beams of beams, most familiar as the frames of trestle bridges and the roof supports within big-box stores. Leaves don't go in for our kind of trusses, but we can readily recognize analogous beam-of-beam elaborations. The leaf lamina itself can combine with veins to form a kind of structure that, while uncommon elsewhere either in nature or in human technology, is neither obscure nor counterintuitive. Inventing a convenient term, I'll call these "virtual beams."

For a start, consider a sheet of corrugated cardboard. It may consist of little more than three sheets of thick, rough paper, but it achieves far greater stiffness than would a single piece of paper of the same weight relative to its area. Its stiffness, of course, comes from its thickness—

its depth, to use the usual term for beams. That ripply layer (sometimes two layers plus a center sheet) does little more than keep the top sheet from getting intimate with the bottom one, but that, with a little glue to keep the layers from sliding around, proves crucial. As with our ordinary trusses, nature has few close examples.

In systems ranging from scallop shells to many large leaves, nature keeps top and bottom separated in another way. This alternative arrangement does without both the top and bottom sheets of corrugated cardboard, in effect stiffening the ripply layer enough so its own top and bottom withstand whatever pushes and pulls might be applied. That's how wavy fiberglass or galvanized steel roofs work—self-stiffening coverings for storage buildings, patios, and so forth. Both we and nature often use a slightly modified version of this wavy, single-layer composition. A sheet of paper can be bent with ease. If it's folded, say, at 90 degrees, then it resists bending in two planes, as in figure 11.4a.

So if we fanfold the paper, taking advantage of the way a fold resists bending, we easily make a properly stiff but lightweight beam or bend-resistant covering. Fanfold roofs came into widespread use in the nineteenth century with the simultaneous advent of inexpensive glass and ferrous metal. The most famous of these was the roof of Joseph Paxton's great Crystal Palace that housed London's Great Exhibition of 1851. Light in another sense of the word, its value for admitting solar radiation must have been a major attraction, with inexpensive electric lights still decades away. Fanfold roofs graced both large railway stations and small greenhouses (these last paying royalties to Paxton) more often in Britain and continental Europe than in North America.[3] "Ridge-and-furrow" or "ridge-and-valley" roofs still get built from time to time, as in figure 11.4b.

A great variety of leaves adopt just this trick, most often if their veins run outward from a central base or in parallel out along a blade. Typically, the veins occur at the leaf folds, both top and bottom (fig. 11.4c). Sometimes a single fold forms a cross section that's either concave upward or V-shaped, putting the thickest part of the leaf where it has to resist compression without buckling—as in a leaf of the ornamental grass of figure 11.4d. Sometimes the leaf simply curves upward at its two sides, with parallel veins distributed across its width—as do a lot of lily leaves.

Both we and nature find such folds and bends a great way to gain stiffness on the cheap. In leaves they also provide a convenient basis for what

FIGURE 11.4. Fanfolding stiffens a planar structure, whether a sheet of thin paper (*a*), a greenhouse roof at Duke University (*b*), or some leaves of false hellebore (*Veratrum*) at Peaks of Otter, in Virginia (*c*). Even a single V-fold will do the job, as in the blade of monkey grass (*Liriope*) in (*d*). I admit that I don't know whether the particular ridge-and-valley roof in (*b*) gains structural advantage from the arrangement—although that was the initial rationale.

engineers call deployable structures, to which we'll return in chapter 13. At the same time, we can immediately recognize a drawback. Folding makes it quite impossible for much of the sunward leaf surface to face the sun directly. Should leaves limit their photosynthetic machinery to one side of each groove and just build some bracing on the other?

No—making both sides equally close to being fully broadside to the sun works better. Consider a leaf with its surface elements bent 30 degrees off the overall plane of the leaf, thus having grooves spread at 60 degrees. It loses remarkably little potential sun exposure with this degree of folding—less than 15 percent. Even with 45-degree folding and 90-degree grooves, the loss is only around 30 percent.* You can do a rough check by folding a sheet of paper or index card and then looking at the resulting reduction in overall shadow area.

Are Trees Columns or Beams?

Many chapters ago, I raised the complementary questions of why trees grow tall and why trees don't grow taller still. This chapter on mechanical support provides a good context for revisiting the business.

We might start by putting the issue in the starkest terms. How tall can a tree made of wood grow before the wood at the base can no longer support the wood above, before it is crushed, as in figure 11.5a? We know the

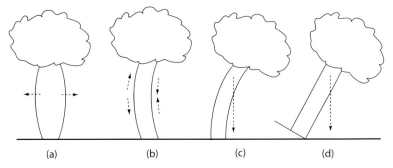

FIGURE 11.5. Four ways the trunk of a tree might fail: (a) simple compressive crushing; (b) Euler bending, in which one side bows outward and the other inward; (c) bending driven by movement of the center of gravity to one side of the tree's attachment to the ground; and (d) uprooting of an otherwise unstrained trunk.

compressive strength of wood. Although most of the tabulated data refer to dried and sliced stuff and aren't exactly applicable, they will be good enough for this crudest of views. Woods have compressive strengths of around 50 megapascals, or 50 meganewtons per square meter. That's stress, not force—that is, the force relative to cross-sectional area needed to provoke crushing. The compressive strengths of woods, incidentally, amount to only half their tensile strengths: woods, wet or dry, resist tension much better than they resist compression. That's unlike, say, steel, with nearly equal strengths, or the other extreme, concrete, which much

*No great mathematical issue—the cosine of 30° is 0.8660; that of 45° is 0.7071.

prefers to be pushed against than pulled on. And we know the density of wood: roughly half that of water, or 500 kilograms per cubic meter. The calculation is simple, the answer absurd.* Crushing becomes hazardous when the tree approaches 10 kilometers high—something over 6 miles, or ten times the height of the Burj Khalifa in Dubai, the world's tallest building. And that's without allowing it any upward taper, which would permit still greater height.

Okay, failure through crushing can be disregarded. Tall, thin, erect structures rarely fail this way anyway, mainly because they face a type of failure that comes into play at less extreme stresses. That's failure by what's called Euler buckling (fig. 11.5*b*), laying the blame, perhaps unfairly, on Leonhard Euler (1707–1783), the great Swiss mathematician who, as one of his lesser accomplishments, derived the formula for it. You can make a column fail under Euler buckling by propping up a short length of dry spaghetti and pushing down on the top. The particularly insidious character of Euler buckling should be apparent: at some critical force the spaghetti begins to buckle sideways. When that happens, it proceeds to break with no additional downward force, indeed even if the force then diminishes.

Assuming that the risk of Euler buckling sets the limit might give a more realistic maximum tree height. To estimate height, we need to specify how the ends of the column behave. Let's treat the bottom as fixed, embedded in unyielding ground, but allow the top to move as if on a hinge—that sounds treelike. We also need to assume a location for the downward loading force—the simplest (but worst) case concentrates the weight at top. The other things that matter are the cross-sectional shape (assumed cylindrical) and the diameter (assumed 1 meter) of the tree and the stiffness and density of its wood. A quick calculation gives a much more believable answer, about 115 meters.† That's about the height of the tallest

---

*First, note that the diameter of the tree is irrelevant—the fatter tree weighs more but also has a greater cross section to support the weight. The mass relative to volume, or density, of 500 kilograms per cubic meter multiplied by gravitational acceleration of 10 meters per second squared gives the downward force relative to volume—5,000 newtons per cubic meter. Divide the crushing strength, 50,000,000 newtons per square meter, by 5,000 newtons per cubic meter to get 10,000 meters, or 10 kilometers.

†The critical force, F_E, for Euler buckling of a long, untapered, circular cylinder with one end embedded is

trees on earth. Recall the assumptions of no taper and the weight at the top—we might reasonably double the number for more realistic conditions. But then an allowance for some safety factor would perhaps undo the doubling—most of the trees of the canopy of a forest could not spend most of their time at the edge of failure.

Yet another approach to the way gravity imposes a mechanical limit on tree height goes back to a classic paper by an English mathematician, Alfred George Greenhill (1847–1927), in 1881.[4] He guessed that the 67-meter (221-foot) flagpole at Kew, made from a single Canadian pine tree, might approach the maximum. He figured that the limit would come when a lateral movement of the top brought it far enough out from the base that the bending would become intolerable and the tree would break, as in figure 11.5c. His result, about 120 meters, assuming no taper, falls in the same range as our estimate from Euler buckling—not surprising, inasmuch as it makes similar assumptions and depends on the same stiffness and density of wood. D'arcy Thompson mentions Greenhill's analysis in the famous second chapter of *On Growth and Form*,[5] so it has retained general attention.

Again, though, reality intrudes. Recall (or repeat) your spaghetti squeezing. Breakage almost always occurs near the middle, rarely if ever near one end. How many fallen trees have you seen that broke well above the ground, as would happen if they failed by Euler buckling or Greenhill's variation of it? For that matter, how often have you seen a broken tree whose wood, judging from the broken end, wasn't rotten? Where I live, that's rare. That mainly happens to middle-sized pine trees during severe icing events, circumstances that add enormously to the gravitational load on a tree.

To me, that limitation to ice storms implies that we (and Greenhill, too) have been looking at the problem the wrong way. Under normal cir-

$$F_E = \frac{\pi^3 E r^4}{2l^2},$$

where E is the stiffness, or Young's modulus of elasticity, of the material; r and l are radius and length of the column. A typical value for E is 6,400 MPa. The load, F_W, is just the density, r, times the volume, pr^2l, times gravitational acceleration, g. Equating the forces $(F_E = F_W)$ and solving for l gives

$$l^3 = \frac{\pi^2 E r^2}{2\rho g}.$$

cumstances, gravity has little relevance in the initiation of toppling of tall trees. (Once a tree starts falling, that is. After failure, gravity becomes inescapably dominant.) If gravity does matter to the tree-height game, it more likely does so, as has been suggested from time to time, through the pressure loss it imposes on rising sap and the resulting risk of embolisms in the conduits.

Where does this reasoning take us? It says that in an engineering sense, these impressively columnar structures, trees, may be beams rather than columns. If they're beams, as in figure 11.5*d*, then they're cantilever beams much like leaves and petioles, except turned 90 degrees. Consider: we say that trees blow over, with "blow" no mere figure of speech. Wind, not weight, provides the force. And the sideways force of the wind must mainly act on the draggiest part of the tree, its crown of leaves. That puts bending stresses on the trunk, as Greenhill suggested. But trunks only rarely fail in the way he (or we, earlier) assumed. In the common mode of failure, trunks are wrenched over at their point of attachment to the ground. So the classical view has two problems. Trees operate as beams and not columns. And as beams they're ordinarily overbuilt, with grabbing the ground a greater problem than avoiding breakage.

What we should worry about, then, is the drag of the crown of a tree. When I walk through our local forest during the winter and look at trees that have fallen during the past year, I notice that most of the corpses (corpses of copses where a local gust did the job) still retain leaves or the remnants of leaves. That tells me they fell when fully leafed out—normal, seasonal loss of leaves begins with specific alterations to their attachments, not leaf death. Conversely, dead trees retain their leaves for many months. And that, in turn, turns our attention back to leaves, now to their drag in high winds, the subject of the next chapter.

AFTER OUR LAST HURRICANE—fortunately a rarity where I live, a hundred miles west of the Atlantic coast—trees failed by the hundreds. Whole copses were leveled, entire hillsides became tangles of trunks and branches. Occasional trees broke, but far more uprooted, perhaps facilitated by some wet weather before the event. A day or so later, one of our local news services contacted me, having learned through some mysterious channel that I had given thought (published some speculations, really) to why trees either do or don't fall over. Could I hazard some statement—beyond an obvious platitude—about the circumstances?

Yes, I could, but not quite what the caller expected. Local winds peaked at perhaps 75 or at most 80 miles per hour (120 to 130 kilometers per hour, 33 to 36 meters per second). As far as we know, few of our trees, long-lived though they are, would ever have experienced winds this strong. Yet faced with that personally unprecedented disaster, relatively few of them blew over. "So few of them failed" was my response to the message about what was newsworthy. Most had apparently been constructed with what a structural engineer would consider an adequate safety factor. Safety factors represent foresight and insurance, neither of them items that easily fit within a simple, perhaps simplistic, view of natural selection.

Not only do trees have long lives, but most go in for growth when young and postpone reproduction for many years. That's only sensible when competition for access to the sky results in winners and losers. An uprooted twenty-year-old that hasn't produced seeds contributes as little to posterity as an acorn eaten by a squirrel. From that point of view, a design with good disaster insurance sounds ideal. Still, the insurance must cost something in terms of growth and seed production, so most of the time the more vulnerable, less well-insured tree will enjoy a reproductive advantage. What matters in an evolutionary sense, it seems, is that more likely than not, such a personally unprecedented disaster has hit at least an occasional one of the tree's forebears—even if not in every generation.

Surviving a Storm

12

Many trees did fail, but we lost no more than a small fraction of the population of any local species. So the situation met our expectation as biologists. If all had failed, then we'd have needed to come up with some special rationalization for the composition of our forest. If none had failed, we would wonder about a counterproductive diversion of resources that might have been better invested in reproduction. In the end, reproductive success is real success, biologically speaking. Naturally (one might say), my response did not make the local paper. Subtlety does not characterize journalism, and scientific journalism provides no exception. A Steve Gould (much missed) could have sold the idea; however prolific and effective a writer, Steve was scientist first and writer only second.* Science writers, alas, are not writing scientists.

Drag

The big mechanical problem for living trees, anticipated in the previous chapter, comes from these severe storms. What damages or destroys them isn't the downward force of gravity on columns but the sideways force of the wind on cantilever beams. While that sideways force, drag, could not be more familiar, it also could not be a less well-behaved physical variable.† It's in a class with convective heat dissipation, if maybe not quite that horribly messy.

Drag matters as well for an automobile or airplane or even a swimmer, but it acts on a tree in a particularly insidious way. Consider levers such as the protruding handles of pruning shears, jar openers, can openers,

---

* Project Steve, somewhat frivolous but with a serious underlying point, memorializes Stephen J. Gould's (1942–2002) effective outreach. Beginning in 2003, people named Steve or any version of the name have been asked by the National Center for Science Education (www.ncseweb.org) to authorize the placement of their names on T-shirts containing a statement asserting the scientific primacy of evolution by natural selection. I'm proud to be in the initial group of 220 names. At that point I was teaching a course with Steven Churchill, who's also on the shirt. My earlier coinstructor, Stephen Wainwright, is on the current version. As of 2010, the number of names exceeded 1100—the current shirt (the "kilosteve" version) is a bit more cluttered than our treasured early editions.

† As an unreconstructed linguistic purist (except for phrases of my own devising), I cringe slightly whenever I run into the redundant phrase "drag force."

or garlic presses. Or think of a lever as simple as a crowbar. A little force (your effort) on the ends yields a much greater force applied to the load, as in figure 12.1*a*. Energy may be a conserved physical variable, but force is

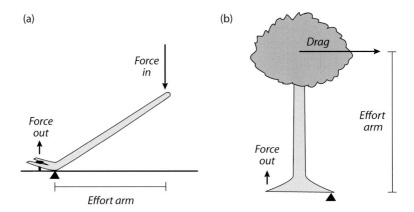

(a)

Force
in

Force
out

Effort arm

(b)

Drag

Effort
arm

Force
out

FIGURE 12.1. (*a*) When you press down on a pry bar, your force gets amplified, because you apply your effort a greater distance from the pivot point than the distance between the pivot and whatever it is you're prying. (*b*) The same thing happens when wind blows on the crown of a tree, prying it out of the ground.

emphatically not conserved—countless familiar devices amplify it. What is a tree in a wind but a giant lever? Drag acts almost entirely on the crown, and the trunk not only conveys the force earthward but amplifies it by some huge factor, enough on occasion to overwhelm its rooting and wrench the tree out of the ground, as in figure 12.1*b*. To make matters worse, the tree has to take soil pretty much as it finds it. Trunks and branches it can build, but soil simply sits there, a material with minimal tensile strength—in other words, with almost no resistance to being pulled against. We'll get back to the problem of grabbing the ground and just how the tree might oppose the wind, but we need to talk first about drag and leverage.

To begin on a formal note, we define drag as a force on an object from the flow of either a gas or a liquid, a force that acts in the same direction as the flow. That distinguishes drag from lift, which acts at a right angle (usually but not necessarily upward) to the flow. (If lift ever matters to leaves, we've yet to recognize the case.) It makes no difference at all whether the object is stationary and the fluid moves across it or whether the object moves through otherwise stationary fluid. And whether the fluid is a gas such as air or a liquid such as water makes no basic difference—quantitatively different, yes; qualitatively, no. Drag follows the density of the fluid. Since water is eight hundred times denser than air, if all else is unchanged, drag in water runs about eight hundred times greater than in air. Drag also follows size, most often but not always tracking some area. And finally, drag follows the speed of flow—but in a less simple way. For ordinary objects in ordinary fluids at ordinary speeds, drag goes up with the square of speed. Double the speed and you quadruple the drag.* That's the main reason you get better gas mileage at nonextreme speeds. If drag varied directly with speed, then it wouldn't matter how fast you went—doubling speed might double drag, but that would be offset by a trip that took half as long.

What about the drag of uncommon objects? The drag of streamlined forms, those with the familiar rounded fronts and rears that taper to points, increases less drastically with speed, about halfway between tracking speed and tracking the square of speed. In addition, streamlined forms have less drag than similar unstreamlined ones at all speeds. But that's another story, since neither trunks nor branches nor leaves seem to employ conventional streamlining. Of more immediate interest, flat plates oriented with the direction of flow—in effect weathervanes—behave like streamlined objects in both respects, with both low drag and a low dependence of drag on speed.

Finally, what about the drag of leaves, the force caused by wind that, amplified by long trunks, makes trees uproot or break? Leaves bear little

---

*We have to add a lot of qualifying words, since these rules of thumb tacitly assume that the Reynolds number (see chapter 6) does not change. And Re, of course, varies with size, density, and speed—just the variables we're looking at. They also assume that the Reynolds number is not too low—very small things going very slowly (such as falling fluff-bearing seeds) follow another rule. The usual variable for expressing drag (D) is something called the drag coefficient, C_d, nothing more than a dedimensionalized drag—drag relative to speed (v), size (S, for surface area), and fluid density, ρ:

$$D = \frac{1}{2} C_d \rho S v^2.$$

C_d can be regarded as an index of dragginess. Choice of S depends on what we want to do with the resulting numbers: for cars (you may have seen numbers for C_d advertised), the area facing flow is the usual choice. For leaves, there's as yet no "usual." I've used original photosynthetic (upper) surface area.

resemblance to the sorts of structures whose drag bothers engineers. In particular, they're flexible surfaces flexibly attached to twigs and branches. One guesses that the most appropriate model for a broad leaf on a tree in a wind would be a flag or banner extending downwind from its pole. We do have some data for the drag of flags, and what's most striking is how draggy they are. To me that's counterintuitive—I'd have thought that being free to follow the irregularities in the local wind would mean that their drag would be especially low. But no: not only is the drag of a flag high, but it increases dramatically with increases in the speed of flow over it—closer to the cube of speed than to the square. Double the speed, and you increase the drag roughly eight times.

Quick back-of-the-envelope calculations suggest that were the leaves on trees as draggy as flags, trees would be far less wind tolerant. And casual observation agrees: in winds, leaves and flags behave quite differently. Flags flutter, often violently, as you can see in especially dramatic form in the flags people attach to their cars to proclaim their infatuation with one or another athletic enterprise. Leaves move around a lot in a storm, but they don't in any obvious way develop the same lengthwise rippling as flags, nor do they often fray along their downwind edges.

Only in a few instances have people mounted trees in wind tunnels, even though tunnels big enough to hold at least the crowns of very large trees have been around since the 1930s. I once stood in the working section of one 30 feet high and 60 feet wide, large enough for a small airplane or helicopter, and it's awesome even with the wind turned off. For an unassailable reason, though, engineers are reluctant to allow trees in their tunnels. To minimize artifacts from the swirl and turbulence of their fans, such tunnels have their fans just downstream from their working section and thus exposed to anything that comes loose. Encounters with chunks of tree flying windward would not be healthy for these huge and undoubtedly hugely expensive propellers.

Still, on occasion, trees have been wind-tested. Back in 1962, a Scotch pine (*Pinus sylvestris*) was subjected to winds from around 10 up to almost 40 meters per second (90 miles or 140 kilometers per hour).[1] Its drag increased steadily as the wind was turned up. Remarkably, drag increased in direct proportion to the speed of the wind, not to the square of the speed. That's a less drastic dependence than for an ordinary object; less, even, than for a streamlined object; and of course vastly less than for a flag in a wind. Even more remarkably, the drag of the tree increased at a slightly lower rate than the drag of a rigid weathervane, even if the actual drag at any given speed (not the rate of increase) didn't quite match that gold standard.

In the 1980s, while writing a short book on how organisms deal with flow, these results tickled my ever-ticklish curiosity. In particular, I wondered how this odd behavior of a whole tree would be reflected in the behavior of its draggiest elements, its leaves. Foresters may care more about whole trees, but I'm in the end a physiologist, scion of a thoroughly reductionist lineage. Until recently, we had a wind tunnel in the department, originally built for Vance Tucker's pioneering work on bird flight. So nothing was easier. No bit of curiosity-driven science that I cared about or was in the middle of could be done with less cost, time, or bother.

Unfortunately, I was working over Thanksgiving break, preparing something for an end-of-the-year symposium on flow, and the local deciduous trees had discarded their leaves.[2] So I tested twigs and saplings of our local loblolly pines (*Pinus taeda*). I got about the same results as had been reported for the whole Scotch pine—drag increased with speed, but only in direct proportion. Twigs of hollies (*Ilex opaca*), each with about seven leaves, gave similar numbers. The pines reacted about as one might have expected, their needles gradually moving closer to the direction of the wind. The holly leaves, though, did something quite surprising. Individual leaves gradually folded down against one another, folding into orientations parallel to the branch. Together they formed a sort of sandwich, with the spurs around their edges seemingly stabilizing the clump, as in figure 12.2. Just as surprising, in a way, the effect was reversible and apparently noninjurious to the leaves—turn off the tunnel, and everything goes back to normal.

With both pines and hollies, the change in drag with speed had an odd feature. Initially, meaning with very low wind, drag increased dramatically with increasing wind. But it never got all that high, simply because at a certain still low speed, things changed in a big way. For the pines at around 5.5 meters per second (12 miles or 20 kilometers per hour) and for the hollies at around 7.5 meters per second (17 miles or 27 kilometers per hour), the increase in drag with speed began to drop off sharply. Yes, drag continued to increase, but it did so more nearly in direct proportion to

FIGURE 12.2. A branch of holly subjected to an increasing series of wind speeds in an especially turbulent wind tunnel. It seems undamaged by the exposure—it immediately regains its original posture when the wind is turned off.

speed—not to the square of speed, as with, say, a baseball, or to the cube of speed, as would a flapping flag.*

Based on these results, I suggested that we ought to avoid applying the word *deformation* to what leaves do in winds. *Deformation* has entirely too pathological a connotation, perhaps the linguistic bias of a culture that makes structures whose function depends critically on rigidity. A less judgmental term, my preference, would be *reconfiguration*.

Five years later I returned to the matter, again as something to play with when the rest of life took a brief holiday. This time, a few weeks in August separated the dispatch of a manuscript to a publisher from the arrival of new students and classes. I made two alterations to the earlier procedure.

*If drag increases directly with speed, then the drag coefficient, proportional to the square of speed, will decrease in inverse proportion to speed: $C_d \propto v^{-1}$. The actual exponents—pine, –1.13; holly, –1.30—are a little lower yet.

First, leaves were no longer tested at speeds too low for drag to have serious consequences. Measuring drag at low speeds was troublesome, so why bother if it didn't matter? And all the testing was now done, not in the usual working section of the wind tunnel, but where the air was blown out into the room, a few feet behind the fan. That gave a higher top speed, almost 30 meters per second (65 miles or 105 kilometers per hour)—quite a severe storm—and vastly more turbulence. By then I had realized that wind tunnels intended for work on flight don't want to simulate natural winds. They're designed to facilitate an odd switch of frame of reference: instead of the craft moving through otherwise still air, the air moves smoothly past a stationary model. So their designers struggle to achieve unnaturally smooth flow, not what a leaf on a tree in a storm would experience.

Sometimes you get lucky. If I had started with single oak leaves, I'd have probably either dropped the whole exercise as a pilot experiment that had an unpromising outcome or gone back to hollies. But my first specimen was the leaf of a tulip poplar (*Liriodendron tulipifera*, not closely related to the true poplars), the very one shown in figure 12.3. As I turned

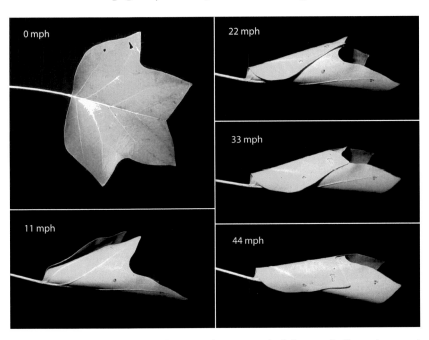

FIGURE 12.3. An analogous reconfiguration happens to single leaves of tulip poplars—and quite a few other kinds of leaves. Again, the reconfiguration is reversible.

up the wind tunnel from 5 to 20 meters per second (11 to 45 miles or 18 to 72 kilometers per hour) the leaf rolled into an increasingly acute cone, with its bottom surface forming the outside. Even in the highly turbulent wind, it scarcely shook and didn't flap at all. After each increment in speed, I walked from the speed control to the leaf and took a picture—without even using a tripod. These "for the record" initial photos have been reprinted quite a few times; they've been the most remunerative photos I've taken in my sixty years of photography. The next day I began attaching leaves to a beam equipped with a strain gauge to measure force. The leaves did about what I had earlier observed with pines and hollies: their drag rose almost directly with speed, not with any higher power (an exponent, nothing theological).* Right away I began getting the data that were eventually published.[3] I've never done easier science.

With that second experiment, I pushed things a bit further. Earlier, I had looked only at how drag varied with speed, but now I measured the area of each individual leaf. A leaf-area meter, borrowed from the local plant ecologists, made the operation even easier than measuring drag. (It's now obsolete—one just uses an ordinary scanner plus a program such the ImageJ freeware from the National Institutes of Health.) With that information I could correct drag for differences in leaf area and make proper comparisons of drag per se, not just how drag varied with speed. It being summer, I had access to lots of different leaves, even when restricting myself to local specimens that could be tested, unwilted, a few minutes after collection. Besides individual leaves, I tested clusters of leaves, rigid plates, and flags. Here's a quick encapsulation of what came out:

- The reconfiguration scheme I found in tulip poplars finds wide use. Among the leaves I tested, cones were also formed by redbud (*Cercis canadensis*), red maple (*Acer rubrum*), sweetgum (*Liquidambar styraciflua*), and sycamore (*Platanus occidentalis*). It has now been reported in catalpa[4] as well. All these leaves have relatively long petioles. In addition, the bases of their blades are lobed and extend a bit back toward the twig beyond the point at which the petiole meets the blade (think heart-shaped, with the attachment at the indent). Those lobed bases

*The exponent for tulip poplar was –1.18. Other leaves turned out to be not quite as low, but almost all had values well below zero, indicating a more favorable speed dependence than a rigid, nonstreamlined body would have.

seem to catch the wind and start the rolling. In any breeze they face up-wind, since the blade trails kitelike behind the petiole.

- In all cases, clusters of leaves as well as individual ones could reconfigure. Clusters formed multileaf conical assemblages with even less drag relative to their overall photosynthetic area than individual leaves had (fig. 12.4). Some leaves reconfigured as clusters even when they did

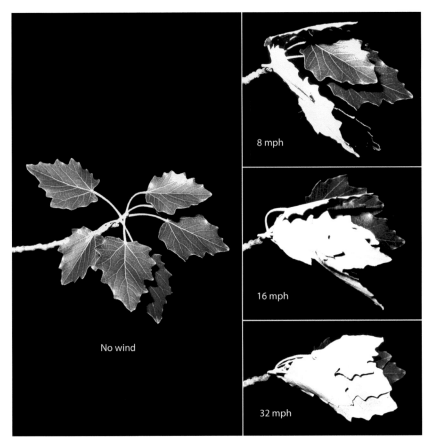

FIGURE 12.4. Yet another reconfigurational series, this one of a leafy branch of white poplar.

nothing noteworthy as individuals—in particular, those of willow oak (*Quercus phellos*), with its long, slender leaves, and the small leaves of white poplar (*Populus alba*).

- Lobed oaks such as the white oak (*Q. alba*) behaved (or misbehaved) in an unusually flag-like manner. Individual leaves flapped badly, with lots of drag and with their drag increasing roughly with the cube of speed. Most suffered some tearing at 20 meters per second (45 miles or 72 kilometers per hour); leaves of other species did not. But when they were allowed to cluster, they formed the same kind of conical groups with about the same drag as did all the others.

- The two kinds of pinnately (pen- or feather-like) compound leaves I tested, black locust (*Robinia pseudoacacia*) and black walnut (*Juglans nigra*), responded to wind differently. Individual leaflets rotated downwind and toward the central rachis (shaft) of the leaf, together forming a tube that decreased in diameter as the wind rose. Relative to original area, they suffered less drag than any of the other leaves. The fronds of an understory palm, *Chamaedorea*, do the same trick with an equally good result.[5] Whole leaves such as those of pignut hickory (*Carya glabra*) that are pinnately arranged on their branches do a clustering reconfiguration of the same kind as these leaflets.
- A leaf-sized square flag has about ten times the drag of leaf-sized rigid weathervanes—flat plates parallel to the wind—at 20 meters per second. Leaves? On average they have less than half the drag of the flag but about four times the drag of the weathervane. Rigidity really pays, but to achieve it a leaf would pay a high price in terms either of investment of material or of sheer weight for the petioles, branches, and trunk to support. Using models of thin flexible sheeting with struts glued on, I tried with only modest success to get at the difference between real leaves and weathervanes. While the exercise told me little about the particulars, it most definitely left me even more impressed with the sophistication of the leaves.

Does all this wind-tunnel behavior actually happen to real leaves on real trees in real storms? Once one has seen one or another form of reconfiguration in the wind tunnel or in photographs, it willingly displays itself during a storm. You just have to look through handheld binoculars while following a leaf or group of leaves as the branches thrash about. But it really helps to have first seen the reconfiguration without all the larger-scale movements.

A year or so after this second experiment, I wrote an article for the magazine *Natural History* on what leaves did in high winds. I suggested that I provide my own photographs, remembering just how easy it had been (and recalling the bother of playing gofer for hurried and harried professionals). But the results, taken in the spring after the end of classes, didn't match what I'd gotten earlier, causing dismal thoughts about science's ever-present risk of self-deception as well as still less agreeable possibilities. Leaves persisted in flapping, shredding, and engaging in other violent, volant misbehavior. Still, because I also persisted, they gradually toed the party—that is, my published—line. Apparently leaves have to harden and toughen for a while after they first expand. I've subsequently noticed lots of leaf shreds on the ground after strong winds in late spring, but fewer as summer comes on. In this peculiarly literal sense, it's all too easy to turn over a new leaf. Again, sometimes one gets lucky. If I'd first played around with leaves in the wind tunnel in the spring, I'd have dropped the project. It does makes me wonder about how often some trivial detail of a pilot experiment may have misled me into thinking that some prescient idea would be experimentally intractable.

Speculations and more questions. When the wind begins to rise before a storm, our local maple leaves flutter and show their silvery undersides. Oak leaves require more wind before they do much moving around. Are we looking at a trade-off, maintaining a photosynthetically optimum skyward orientation up to higher speeds (oaks) versus better reconfigurational ability (maples)? And is the especially dense, strong wood of oaks part of the price paid for the better productivity in breezy habitats? That's what I take away after playing with white poplars. The genus *Populus* includes several trees that often feel high winds: cottonwoods, quaking aspens, white poplars, and others. The leaves of all share an attractive shimmering behavior in gentle breezes. Is that shimmering ("quaking") no more than the low-speed fluid-mechanical instability that accompanies good reconfigurational ability in severe winds? In other words, is it a more dramatic version of what maple leaves do?

Pinnately compound leaves are especially common among the canopy trees of tropical rain forests.[6] These trees tend to be skinny as well as tall, and strong windstorms such as hurricanes occur while the trees are fully leafed out. If we include palms—with much longer leaflets—as well, based on the reconfiguration of the relatively diminutive palm *Chamaedorea*, that tropically biased distribution of compound leaves becomes still more pronounced. Does their especially good reconfigurational ability and consequent low drag in high winds give them a competitive advantage where they live?

Do it yourself: While wind tunnels aren't as ubiquitous as leaves, a wind machine a lot smaller and less powerful than the one in my department would have done as well for my project. So lack of a wind tunnel shouldn't stop a person from making wind. Attach a leaf—most petioles have a convenient thickening where they meet their stems—with string or wire to a strut. Then stick the leaf out the window (passenger's side, please) of an automobile going down a road and, if you want, take pictures from inside it.

Or get a long piece (5 or 6 feet long) of flat wooden molding. (Mine happens to be edge molding $\frac{3}{16} \times \frac{5}{8}$ of an inch in cross section.) Make a small hole near the end for string or wire, and attach a leaf or cluster of leaves. Then, treating the molding as an extension of one of your arms, rotate your body, and look out along your arm as you turn faster, as I do in figure 12.5. Moving the leaf as fast as 9 meters per second (20 miles or 30 kilometers per hour) is no problem—at least until you get too dizzy to stand.

FIGURE 12.5. The author (in the mid-'90s) demonstrates how we can view reconfiguration without needing access to a wind tunnel. The downside, of course, is dizziness. Mike May took the picture in connection with a spread in *Muse*, a science magazine for children (see note 7).

Where are we now? No current theory enables us to look at a leaf and make a decent estimate either of its drag or of how drag varies with wind speed—although I know of one promising investigation currently under way. Nor do we have much data to back up any of the generalizations I've been bandying about. How reorientation of the branches of trees affects the overall picture remains obscure. Lots of additional issues rattle around in that limbo of unasked questions, largely, I'd guess, from simple lack of attention. Science, like other human activities, tends to proceed along lines set by trends and fashions and in directions determined by potential profit. In our real world, this last must be measured by availability of financial support as much as by intellectual or practical gain.

Leverage and Anchorage

Several threads now converge—as anticipated when drag first made its appearance. First, as we concluded in chapter 11, as far as mechanical failure of their trunks goes, most healthy trees most of the time are overbuilt. In words of one syllable, they don't break. Leaves in wind pull a tree windward, and via the long lever arm of the trunk, that drag tries to wrench the tree from the ground. Leverage—that has to take center stage. A tree doesn't get pushed over as much as it gets turned—in the end by 90 degrees.

What matters, then, to put things in proper terms, isn't force itself but what's called turning moment, the force (here drag) multiplied by the shortest distance between the line of action of the force and the pivot about which the system rotates (effort arm or moment arm), as we saw earlier, in figure 12.1.* Forces, properly, can't be properly specified by single numbers the way masses can. A force really has three parts—its

---

*Turning moment, otherwise known as torque, thus has dimensions of force times distance. These are the same dimensions as those of work and energy, but don't make the mistake of equating them. Critically, here the directions of force and distance are at right angles to each other rather than being in the same direction and on the same line. The units of turning moment or torque are newton-meters in SI or foot-pounds in the American vernacular. A turning moment not balanced by another one of equal magnitude and opposite direction will cause angular acceleration—turning—of any object with mass, just the way a force not balanced by another force will cause a mass to accelerate along a line.

magnitude (the number you unusually see), its direction, and its line of action.* For the drag on the crown of a tree, the direction is that of the wind, since drag is always a force in the direction of flow, and the line of action is where that drag acts, here well above the ground.

In designing a tree that will remain erect, perhaps the greatest challenge for nature is neither minimizing the drag of leaves nor making an adequately strong trunk. Somehow the tree has to grab the ground, peculiar stuff over which it has only a little more control than it does over the properties of the wind. Ground ordinarily has precious little tensile strength, so when pulled on, it fails all too easily. By contrast, it resists downward pushing fairly well and sideways pushing only a little less effectively. Only with extensively ramifying roots or with a general root tangle from other trees can it withstand a hard pull. After you free the trunkward end of a substantial root, you're not likely to be able to pull it straight out lengthwise—you have a better chance pulling it upward. So a root in the ground will take some tension. But your maximum pulling force, less than 40 kilograms (400 newtons, properly, or 100 pounds), approaches complete irrelevance next to what a good wind on a high crown of leaves can muster.

To remain erect when provoked by a storm, a tree must provide a compensatory turning moment. That's one of at least equal size (force times distance) and with the opposite direction of rotation from the moment imposed by the drag of its leaves and the length of its trunk. That means it needs some combination of force and leverage, one multiplied by the other. Force amounts to the push or pull before the soil gives way. Leverage depends on where the tree pivots as it turns—or would turn, if its anchorage fails. If a tree can arrange a longer lever, it will need less force, or

*We refer to quantities whose specification requires both magnitude and direction as "vector" quantities, distinguishing them from mere "scalars" that have only magnitude. Combining vectors has to take direction into account: if two ropes pull on an object, their combined force will obviously depend on whether or not they pull in the same direction—and, if not, on the angle between them. Velocity, momentum, acceleration, force, and some other quantities are vectors. Energy (in any form), time, and mass are scalars. Combining a vector with a scalar, as in force equals mass times acceleration, presents no problem—the resultant vector (here force) is the same as the input vector (acceleration). Complications enter the picture when more than one vector is involved.

with more force, a shorter lever will do. Naturally, roots have to ramify if they're to provide water and dissolved minerals, so some degree of anchorage comes with the territory. Still, we've no reason to suppose—and lots of reason to doubt—that a design optimal for getting material into a tree will coincide with one optimum for anchorage.

What options might be open to a tree? Back in 1995 I was asked, on short notice, to give a talk for a gathering of arboriculturists—urban foresters—presumably after a more desirable speaker had ducked out. That prompted a hasty literature search (pre-web) on windfall (or "wind throw") in trees. To my surprise, I found relatively little beyond some accounts (and counts) of what fell and where. So, feeling saved—finding no literature constituted a windfall of a different kind, a license to speculate—I tried to imagine possible schemes. Then, using the literature on root structures, I tried to guess which trees might be using which to stay upright. The audience of very practical folks received the talk better than I hoped. They told me they had long been frustrated that so little good science was around to help them decide what might break, what might fall, and how to prune. So maybe I, not practicing arboriculturist but academic biomechanic, might be a harbinger of help. Since then quite a lot of good work has been done, some of the best by Roland Ennos and his associates, at Manchester University (UK).[8] Still, even now we've only scratched the surface and have yet to dig decently deep.

The commonest tall trees where I now live are pines (loblollies, *Pinus taeda*) and oaks (Southern red oak and white oak, *Quercus falcata* and *Q. alba*). I was familiar with such oaks from my youth in upstate New York, but a front yard of tall pines provided a good scare when the first thunderstorm hit. Big oaks stand rigidly, but pines, even very large ones, sway alarmingly. Obviously, not all trees deal with the anchorage problem the same way. So what are the options, and who might do what?

Take a large, stiff-trunked tree, such as one of those big oaks. We know the components of the uprooting moment: the drag on the crown and the distance from the center of the crown to the ground. What, then, are the components of the anti-uprooting, anchorage-securing moment? The best hint comes from a peculiarity of an uprooted tree. Most often, a wide footplate of roots has sufficient stiffness to resist crushing and buckling well enough to hold the trunk clear of the ground, as in figure 7.7. Between its substantial diameter and its particularly dense wood, that trunk can weigh several tons. On inspection, one notices that the root plate, now upright, isn't all that thick and the hole it left isn't all that deep. Only a little less obviously, the original position of the upright trunk had to be some distance from where the plate now rises from the ground. It looks as if the whole system rotated about a point considerably to one side of where the trunk had originally met the ground, as in figure 12.6*a*.

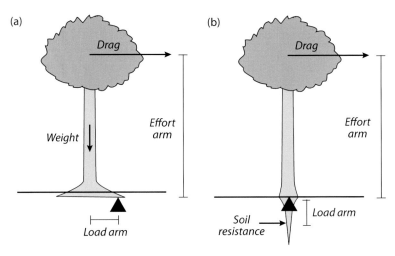

FIGURE 12.6. (*a*) A leverage diagram for an idealized oak-type tree in a wind. (*b*) A similar diagram for a pine-type tree.

Thus we can recognize the main components of the anti-uprooting moment. These are the downward force of the tree's weight (its mass times the earth's gravitational acceleration) and the moment arm as the sideways distance between the trunk and the point at which it pivots as it uproots. The tree depends on having a heavy trunk and a wide, stiff base. The wider that stiff base, the more stable the tree—even if it can't grab the ground more than the little bit needed to keep from sliding sideways. Soil piled on the footplate provides just that much more weight, that much more that must be lifted for failure to occur. Soil held beneath the footplate by short and insubstantial-looking vertical "striker roots," as on the left in figure 12.7, adds still more weight. That's one approach a tree can take—we might call it the oak scheme.

For staying upright that way, a swaying trunk would be decidedly counterproductive. That's because moving the trunk's weight over toward the pivot point would reduce its effectiveness by reducing the moment

FIGURE 12.7. On the left, striker roots emerge from the base of a large oak that had fallen about a month before the picture was taken. On the right, the basal part of the taproot of a large pine that had fallen some years before the picture was taken, wrapped with tape for photographic contrast.

arm. So the center of gravity* of the aboveground mass of the tree must be kept in place, which demands that the trunk be stiff as well as heavy if the oak scheme is to work. Obviously, those swaying loblolly pines in my front yard, with their narrower trunks and less dense wood, must be play-

---

*"Center of gravity" is a wonderfully useful fiction. It's a point at which the mass of the object acts as if concentrated—infinite density at an infinitesimal point instead of a mass distributed throughout the object. If you hang the object from a string, the downward projection of the string will go through the center of gravity. Hanging the object from several points in sequence allows you to determine the center of gravity in two or three dimensions—all downward projections from the hanging point will go through one and only one point, the object's center of gravity. One caution: center of gravity isn't adequate for problems involving rotation, so once a tree starts to fall, things get messier. For the same reason, balancing an automobile tire properly demands not that its center of gravity coincide with the middle of its mount (static balance) but that a slightly different variable, moment of inertia ("dynamic balance"), do so.

ing a different game. They certainly fail differently—when they fall, their trunks most often lie directly on the ground. As it happens, a tall pine has something that an oak lacks: a substantial root descending just beneath the trunk, a so-called taproot, one of which is shown on the right in figure 12.7. Since it more often than not breaks off when a pine falls over, it's easy to overlook. I once obtained an intact taproot. It had pulled out as the tree fell during a hurricane, helped by the sponginess of the ground, which was already fairly well soaked when the storm hit. The root formed a long cone, about 10 centimeters (4 inches) in diameter where I cut it, and both its high stiffness and density made it feel like no other piece of pine.

So what provides the anti-uprooting moment for these pines? It looks as though for the force component they take advantage of the way soil resists a sideways push. If the bottom of the trunk provides a pivot point, as it appears to do, judging from the position of trees that have fallen, then motion of the trunk to one side forces the taproot to move sideways in the other direction. A long, stiff taproot pushing against dense, unyielding soil can then keep the tree erect—even though the trunk sways, as in figure 12.6b. The push of taproot against soil thus provides most of the force in the anti-uprooting moment, while the length (average, more or less) of the taproot is the length of the lever.[9] What of the weight of the trunk? Swaying moves the center of gravity sideways, so weight shifts to the negative side of the ledger—by contrast with the trunk of an oak, it's part of the problem, not part of the solution. A low-density trunk should be best. So that's the pine's scheme, or so it appears.

These two arrangements aren't the only ways trees might remain standing in a storm. Typical tall trees of tropical rain forests clearly can't work the same way as either oaks or pines. Their thin trunks sway in a wind, their root plates may be wide but they're not especially stiff, and they lack taproots. What they do have, most often, are thin but wide buttresses extending downward and outward from the lower parts of their trunks—as in figure 12.8a. They seem to make use of the general tangle of roots in the top layer of soil on the forest floor, taking advantage of soil that can, as a result of that reinforcing tangle, withstand tension. Blow on a tree, and its upwind roots can resist being pulled lengthwise. But the junction of the thin trunk and thin root plate represents a real liability.

FIGURE 12.8. Diagonal supporting elements. (*a*) shows a particularly impressive buttress on a tree at the Smithsonian station on Barro Colorado Island, Panama. The person, of normal stature, is Robert Dudley, now (twenty or so years later) professor at the University of California, Berkeley. (*b*), the Duke University Chapel represents our ordinary sense of architectural buttresses—although we more commonly comment on ones that contact the main structure only near the top, so-called flying buttresses. These take compressive loading, not the tensile loading more likely for the buttresses of trees. (*c*) shows nothing more than a wire stabilizing a utility pole, working in a way closer to what happens in the buttress of a tree.

That's where the buttresses come in. We know buttresses mainly as components of Gothic cathedrals. That analogy misleads, since massive architectural buttresses (figure 12.8*b*) resist the compressive loads imposed by walls that are pushed outward by the downward and outward splaying of peaked roofs. The side-to-side thinness of the buttresses on tropical trees means that they can't resist compression very well—anything beyond a slight downward push will make them buckle. They must act mainly as guy ropes, diagonals running downward and outward from trunks that transmit forces from swaying trunks out to the root plate. Thus the trunk-root junction needn't take the whole load. Why don't such guy ropes look like ordinary ropes or perhaps vines? Probably because as trees grow larger, they grow by outer expansion without reabsorbing less useful inner portions. Thus pines, for which heavy trunks should be counterproductive, remain nonhollow, retaining a core of wood as an odd consequence of the way they grow.

Better analogs than Gothic cathedrals are all the objects we keep reliably erect with guying cables—antennas of all kinds and even utility poles, especially those from which wires don't run in opposite directions (figure 12.8c)—and, on a large scale, cable-stayed bridges. The wide bases of big oaks act as buttresses working in the familiar compression-resisting way, but they're partly buried and not so conspicuous—or maybe they're too contemptuously familiar to merit a common name. Tropical trees have, in addition to the aboveground buttresses, an underground counterpart, an array of striker roots penetrating vertically, like those of the oak in figure 12.7. Each part, then, contributes to keeping tall, slender, shallowly rooted trees firmly anchored to the ground.[10]

Bamboos seem to stay upright with yet another game. Strictly speaking, their stalks aren't proper trunks—bamboos have culms instead—but mechanical problems know nothing about the niceties of botanical nomenclature. Unlike most of our familiar trees, bamboos (like palms) don't expand in girth as they grow. Their culms aren't unbending and heavy, nor do they have much in the way of taproots. I dug up one about 10 meters (30 feet) high and worked my way into it with clippers and water jet, taking pictures every so often. Several features drew my attention. When I pruned away a lot of superficial stuff, what remained was a remarkably unyielding sphere of branching rootlets and dirt just below the culm and ground level (figure 12.9, left).

Trimming further, all but a single passing root (rhizome) came away, since these other roots lacked any direct mechanical or vascular link to the plant. And the connection between the parent rhizome and the base of the culm (the latter far thicker than the upper parts of the tree) turned out to be remarkably small and of no great mechanical robustness (fig. 12.9, right). Bamboos seem mainly stabilized (I'm guessing) by the resistance of that sphere of rootlets to rotation in the soil, aided by the rhizomes that penetrate and run outward from the sphere—whether or not they're biologically connected.

Those rhizomes are nicely designed to pull against the soil, since they have rootlets extending outward at every node, and the nodes are only a few inches apart. And the rhizomes extend a long ways from the culms. After buying a yard that hosts a patch of the spreading rather than the clumping kind of bamboo, I learned the hard way that the rhizomes are both horribly invasive and exhaustingly difficult to dig out. Worse, these

FIGURE 12.9. The bottom of a bamboo, no easy thing to dig out, incidentally. On the top is the tangle of roots after a thorough wash with a jet of water. Most of the large radial rhizomes simply pass through the tangle—on the bottom is an enlarged view of a lengthwise cut (radial section to a botanist, midsagittal to a zoologist), made in an effort to locate the actual connection (*arrow*) of the culm to the larger rhizome.

wonderfully tough lateral rhizomes sprout fast-growing culms in all the wrong places.*

*I hasten to thank Jingjing and Jiahua Xie, who every few days each spring harvest the newly emerged shoots and thus help keep the bamboo from further encroachment on what we regard as our domain, not theirs.

So—compressive buttressing, as in large oaks; taprooting, as in large pines; tensile buttressing, as in many rain forest trees; and probably resistance to shear in rotation, as in bamboos. But this simple list does need some complicating comments. While some schemes appear to preclude simultaneous use of others, a tree can, and many probably do, combine elements of several. Even more commonly, a tree can depend on different anchorage systems as it grows larger. Compressive buttressing works best if the tree is large, so many of its users go through a taprooted earlier. And taprooting does find use in some rain forest trees.

The rarity of a solution that would occur to any structural engineer, one that for us humans has long been the default, points to a serious constraint on most of nature's structures. Only for some special purpose would we support a wide structure above a single column. That a tree has a single trunk is a bit odd, at least if one happens to think about something too familiar to ordinarily draw attention. Does only the advantage of terrestrial mobility elicit the evolution of multiple supports such as legs? After all, sea anemones, sea palms, sea pens, and other erect, sessile organisms depend similarly on single supports.

At least one large tree does make multiple supports: the banyan (*Ficus benghalensis*) of Southeast Asia. A banyan drops aerial roots from its spreading branches, and these mature into secondary trunks (fig. 12.10). Near Bangalore, India, I once saw an individual tree that covered more than an acre—only from the air would it have been possible to take a picture of the whole thing. So it can be done, and a tree can become its own forest. But adding a secondary trunk may be peculiarly difficult for a tree. An aerial root that has rooted must be alternately pulled on and pushed against if the original trunk sways even a tiny bit. So it must either be stretchy or be tolerant of buckling. Both of these attributes, though, are inappropriate for a trunk that provides support as a compression-resisting column. Fortunately for trees, woods do vary a lot in their mechanical properties. Thus the wood of the roots of a pine tree is much less stiff than the wood of its trunk. But how does a tree make wood that works one way and then convert it to wood that works in an antithetical way—without disfunctionality in the transition?

We're ignoring many other relevant matters in this simple, probably simplistic, view. High winds near the surface of the earth come in gusts, and all trees do some amount of swaying. So uprooting really needs analy-

FIGURE 12.10. A banyan tree. This one graces (to my eyes, they're really quite ungraceful) the campus of the Indian Institute of Technology, Chennai.

ses that go beyond some simple computations based on turning moments that might be measured by winching over trees.[11] The combination of the gusting of the wind, moment-to-moment leaf shape change, elastic swaying, damping of the swaying, and vortex shedding by trunks adds up to a suite of mechanical problems that would give an engineer nightmares.

EACH SPRING, the tree-lined street in front of my house disappears from view, not to reveal itself again until late autumn. So swift is that yearly transformation that even day-to-day changes catch our attention. Buds that gradually developed over the previous months burst open as leaves emerge, some individually, some in clusters, and some in profusion along a rapidly elongating stem. So familiar is the annual event that, however conspicuous and aesthetically gratifying, we ignore its particulars: the transformation of tiny individual buds into broad leaves. Even at its speediest, the emergence of a leaf plays out in slow motion, changing at too leisurely a pace for nervous systems tuned to events that demand immediate responses.

Making Buds into Leaves

Leaves don't come from leaf factories at all similar to the assembly lines that generate our industrial products. Nor are they unfolded and extended with help from temporary formwork, molds, or scaffolding. Ultimately, they respond to genetic instructions that direct biochemical syntheses of fabulous specificity. But almost all that happens well before buds become leaves. The visible process consists almost entirely of deployment. *Deployment*—its military application refers to the placement of one's forces, with the presumption that recruitment and training have happened earlier. So the term fits. Even better, it has been applied to a specific human technology closely resembling leaf emergence: the design of deployable structures. Still better, the parent Latin, *displicare*, means "to unfold," much closer to what leaves do than to soldiers arraying for battle.

One deployable structure could not be more familiar or provide a more accurate exemplar. That's the umbrella—folded into a compact package until needed and then deployed with ease, speed, and far less information than was required for its construction. Quite a different version, and almost as common, is the metal tape measure. We store it rolled flat in its container and then draw it out for use, whereupon, by elastic recoil, it changes its cross section from flat to a more bend-resistant arc of a circle. As soon as pack animals came into use—perhaps even earlier—ancient nomadic humans began living in deployable dwellings. With time, deployable structures have taken on ever more diverse roles. Just look around your house at the diversity of devices that become functional once unfolded, extended, or inflated: shower curtains, canvas shopping bags, paper-towel racks, portable drying racks, rabbit-ear antennas, stepladders, and more. They've become ever more prominent in aerospace technology, from simple parachutes almost a century ago to all manner of hardware that can be put into orbit and then deployed into functionality with minimal manual manipulation.

At the same time, we've come to appreciate how much the development of organisms depends on deployment of self-assembling components. Genes may direct the assembly of specific sequences of amino acids into proteins, but after that systems—subcellular, cellular, and multicellular—go together with little further informational input. All they need are appropriate media or substrata. Back in 1950, Horace Crane, a physicist, presciently predicted that living systems would make heavy use of geometries favorable for self-assembly: helices, geodesic spheres, and so forth.[1] A decade later, David Raup, a paleontologist, showed that simple computer programs could generate practically every kind of mollusk shell. All such shells come in geometries that share a particular peculiarity. They can increase in size by adding material to their edges and surfaces without changing their overall shapes—in effect, continuous and informationally trivial deployment.[2]

The emergence of leaves from their buds must be the most conspicuous application of deployable structures in nature. While a little synthesis of material may go on, most of what one sees involves unfolding, unrolling, inflating, and then hardening. By plant standards, the processes proceed rapidly, although the final hardening can extend for weeks beyond fixation of the basic size and shape—the previous chapter noted how early in the season leaves tear more easily in moderate winds. In putting together this account, I've been surprised at how little attention the deployment process has received compared with the vast volume of work on the information-dependent earlier events in development.

We have to be impressed with the deployment process, at once awesome and attractive. Figure 13.1 consists of photographs I took a few days apart as a bud of a mockernut hickory (*Carya tomentosa*) sapling became a group of leaves. From practically nowhere—a small bud—several com-

FIGURE 13.1. Over a period of about twenty spring days, the bud of a hickory develops into a recognizable set of compound leaves. The scale differs among the images.

pound leaves, each with its seven leaflets, detach, unwrap, spread, and harden.

The nonwoody (herbaceous) parts of plants have a remarkable ability to change their mechanical properties. The most obvious, best understood, and undoubtedly most important driver consists of changing the volume of water in particular cells. Adjustment of osmotic status alters cells' internal pressure and therefore volume; doing this within a patch of cells can spread a membrane, cause a leaf to wilt downward, and change their shape in still other ways.

My favorite example of a local change in mechanical properties isn't an unfolding leaf but the bending of the stems of daffodils just beneath their yet-to-unfold flowers. Initially, the flower bud points upward, but in the course of less than a day part of its stem softens and swings downward in an arc of between about 60 and 150 degrees. Thereafter, the softened patch stiffens, and the flower retains that new orientation. Rather than requiring any active push, in this instance gravity provides the flower's force. While a casual look at the specific drooping angles of flowers certainly points the finger of suspicion at gravity, reversing gravity provides a more convincing demonstration. A flower that's allowed to mature and emerge while facing downward rather than upward just doesn't bend over. Reoriented upward, its long axis just continues that of its stem, with the flower facing expectantly upward rather than shyly downward—as in figure 13.2.

FIGURE 13.2. Two daffodil flowers. The one on the left was picked after it opened. By contrast, I picked the one on the right while the bud remained vertical and mounted it tip downward in a rig, as shown in the next figure. Once fully formed, it was reoriented right side up.

Still, gravity isn't the only agent, even in this instance. As mentioned earlier, osmotic adjustments of the water content of cells provide a proper engine. They convert the chemical energy expended in moving around

Do it yourself: A siphon of sorts is all you need to make a flower emerge on an inverted stem. As in figure 13.3, arrange a narrow-neck bottle, a piece of flexible tubing, and a seal of absorbent tissue or cotton so that a freshly cut stem points downward.

Raise or lower the bottle until the pressure across the seal is just enough so that the junction neither leaks water nor draws in air. The flower can be removed as soon as it has started to expand—the stem quickly hardens enough so it won't droop over. Make several of these rigs, and you have a good attention-getter and conversation piece.

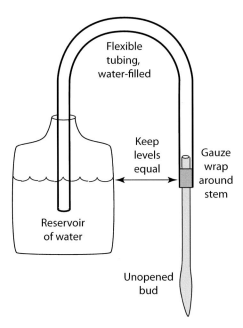

FIGURE 13.3. The rig for persuading a daffodil flower to emerge facing upward and extending symmetrically above its stem. It should be set up where initial drips and leaks won't cause trouble.

osmotically active molecules into mechanical work—making something move some distance against an opposing force. The deploying leaves of figure 13.1 droop downward and then rise and stretch outward—gravity opposes that rise, so the plant must be doing mechanical work. (It probably takes longer than you want to wait, but if you get the variables right, you can make a tree lift a weight, a more impressive motor performance. Just bend a branch downward and insert it into a pot of cement or plaster just heavy enough to contact the ground; the branch will slowly lift the weight.)

The deployment of the hickory leaves in the figure provides only a general model. Some leaves produce persuasively similar miniature versions of themselves that then undergo an overall expansion. The white oak (*Quercus alba*) leaves of figure 13.4 do that. I've seen no good account of

FIGURE 13.4. Emergence of some leaves of white oak over ten days. They mainly expand, so to show how little shape change occurs, I've adjusted the scale so that each successive photo shows a less magnified view.

just what that expansion entails—what mix of cell division, cell enlargement by intake of water and internal chemistry, shifting of cellulose microfibers, and special syntheses of extracellular material.

An investigation a few years ago focused on the topology of leaf deployment. Unusually, it aimed at providing a basis for biomimetic deployable structures—in effect, self-organizing systems. Even less usually, it drew on origami, on a pattern called Miura-Ori in particular. (You'll not be surprised that the first author of the primary paper was Japanese.)[3] The particular leaves studied, those of hornbeams (*Carpinus*) and beech (*Fagus*), are pointed ovals in shape (or, as the botany books put it, lenticular), with parallel lateral veins branching at an angle of around 60 degrees from a midrib, as on the left in figure 13.5. Here we can skip over the in-

FIGURE 13.5. Leaves of an American beech (*Fagus grandifolia*) on the left, with the origami version on the right, unfolded above and held collapsed below.

vestigation's formalities, ones befitting a paper published by the Royal Society of London, and jump to the appendix, which all too briefly (for nonorigamists) describes a deployable folded-paper model. On the right in the figure is my version, expanded and collapsed, produced from the instructions in the paper.

Just as the art of origami has much in common with the deployment of leaves, the two together have a curious connection to contemporary technology. For aesthetic, mechanical, and aerodynamic reasons, we prefer automobiles with rounded body panels. But we make them by stamping sheet metal—two-dimensional planar stock. So only shapes that take kindly to this slight three-dimensionalization can be incorporated into anything except really pricey hand-crafted vehicles.[4] What we do with giant stamping presses differs from origami only in that they produce curves rather than sharp folds from the starting stock. Fiberglass and

More do-it-yourself projects: The following is a more detailed set of instructions for Kobayashi's Miura-Ori model. Cut a piece of ordinary typing paper so that the longer side is twice the length of the shorter side. Fold and flatten it down the middle lengthwise. Now mark on the paper, at intervals of half an inch or a centimeter, a set of parallel diagonal lines at about a 60-degree angle to the fold. Fanfold the paper along those lines to make distinct creases. Open the original fold; you should have a lot of ridges and valleys. Cut diagonally from the center outward at top and bottom at roughly the same 60-degree angle, to get a six-sided figure. You'll notice that the ridges on one side become valleys on the other, and that has to be fixed. So (with the necessary sleight of hand) convert ridges to valleys and valleys to ridges on one side of the original center fold. Voilà—a persuasively leaflike, nicely stiffened model. But that's not the best part. The whole thing can be pressed flat into an undeployed structure.

other composite materials are molded rather than stamped, so one supposes that the shift, slowly happening, to these materials would remove a constraint on designers and lead to now-impractical forms. I don't notice that happening. On a visit to the General Motors research lab (to give a talk), I asked why the nice plastic panels of our Saturn weren't more common among their cars. I was told that the designers didn't like them, because they required a 3- rather than a 2-millimeter gap where doors, trunk lids, and such fitted in. Form your own opinion about priorities.

Heading Off Herbivores

In an all-too-immediate way, animals parasitize plants. Not that the small number that live as pure carnivores can claim innocence—they just outsource their herbivory. Our garden suffers the depredations of rabbits and deer. Both have an inordinate preference for garden foliage rather than the wild stuff beyond our property line. To me, that suggests that evolution has wrought just what might have been expected as we herbivorous humans select plants for palatability. Or maybe herbivory just leaves less evidence in a fast-growing, diverse, and irregular thicket. Still, except when we raise and pasture them, mammals don't amount to more than minor nibblers of leaves, and birds avoid leaves almost completely—for a reason I'll get to in a few pages. Herbivorous insects pose the real threat.

By summer's end, insects will have eaten some of nearly every leaf, but most of the time they leave food on the plate, as it were—odd, when one thinks of it, that a leaf that's clearly edible does not get fully consumed. The only nonarcane explanation I can think of invokes the limited mobility of larval insects. A flying adult lays eggs on or near a leaf, and then each larva eats just enough to bring itself to pupation, never more. Does an adult female mete out her eggs so that leaves don't get fully consumed, rather than, as one might say, putting all her eggs in one basket? In the short run, that dispersal prevents disaster: a large number of simultaneously hatched larvae running out of food before they reached the critical size that, when sensed, triggers pupation.[5] In the long run, it maintains the resource. A tree can survive one or perhaps two seasons of full defoliation, but that's about the limit. Of course, any severe defoliation puts it, even if briefly, at a competitive disadvantage.

Leaves resist the worst consequences of the insects' puncture holes. If you cut a large nerve, everything downstream is effectively denervated. If you occlude a large blood vessel, you make trouble for tissue downstream, although if you have some residual circulation, then over time bypassing (collateral) blood vessels may develop or enlarge. Most leaves don't suffer any equivalent loss of function. The ends of cut veins somehow plug up, and most veins form parts of reticula, or networks, rather than of purely branching arrays. So fluid flowing in either direction can go around a hole. The reticulum of veins has an obvious role in mechanical support, but this failsafe transport benefit hasn't received much attention. At least not until recently. A group of biophysicists noted that an intact and unthreatened optimal distribution system has the geometry of a branching tree, not a reticulum. They then looked at how looping pathways allowed portions of leaves beyond an interrupted vein to continue to function.[6] I tried punching holes in some leaves, and I found, as they did, that the holes had surprisingly little effect. You can see in figure 13.6 how dyed water still flowed everywhere after hole punching.

If leaves were animals, they might hide whenever those who would

FIGURE 13.6. Two leaves of the appropriately named red mulberry (*Morus rubra*). The one on the left has had its petiole dipped for a day in a 1-to-1 solution of dye intended for coloring fabric: Rit liquid dye, wine color. The inset shows the region of the hole at a greater magnification. The absence of dye right around the hole suggests either plugging or constriction of sliced vessels, but I've run into no specific mention of the phenomenon. The color has been intensified for the photograph.

feast on them were in the area. Immobile structures have no such option. Nonetheless, leaves don't just stand there and endure their own gradual destruction, even if they do just stand there like fancy photosynthetic bumps on a log. They've evolved a wide array of deterrents, an array that's amazing whether we consider general strategic categories or specific tactical details. They deter in at least three general ways, with no one tactic excluding the simultaneous use of any other. First, a plant can do its best to poison potential herbivores with some evil chemical. Second, it can build leaves that present physical problems, such as being hard on jaws, mouths, tongues, or teeth. And third, it can make leaves that simply cost too much to process relative to their yield, whether the expense comes in biting, chewing, or digesting.[7]

Whatever the specifics, leaves that cost the plant more to build and maintain or that will be retained on the plant for longer—more valuable leaves—generally elicit more elaborate defenses than do cheap, more expendable ones. The leaves (and other photosynthetic structures) of desert plants may look like nice, succulent morsels—we do call them succulents, which makes them sound toothsome—but they tend to be the most elaborately defended.[8] Less common among plants are structures designed to propitiate hungry animals, devices such as the expendable tails of some lizards.

First, then, chemical warfare, since even an account that tries to keep chemistry at arm's length ought to say a few words about poisons in plants. To be most effective, toxicity should have two actions, although nothing says that the same chemical can't perform both. The more obvious is the poison itself. But in addition, the potential herbivore ought to be deterred before consuming a bellyful of toxin. That's a role either for something seriously unpalatable or something with a characteristic taste that the herbivore has learned (personally or evolutionarily) to associate with digestive distress.

Whatever the tactics, the warfare never ends. Some leaves put their herbivore deterrents in various hairlike outgrowths—trichomes—ensuring that tiny herbivores will encounter these first. Animals figure out which parts of a plant are toxic and then avoid them. Or they evolve detoxification or toxin sequestration systems. In some extreme cases, animals appropriate the toxins, storing them to make themselves toxic. The best-known case is that of monarch butterflies and their host plants, milkweeds: borrowed toxins make the butterflies unpalatable to insectivorous birds. The evolution of specific detoxification or toleration mechanisms by specific animals explains a lot of the extreme specificity of many plant-insect feeding arrangements. We get hornworms on our tomatoes but on nothing else—around here they like tomatoes and tobacco, both in the family Solanaceae. We humans avoid almost all members of that famously toxic family; tomatoes and potatoes are rare edible exceptions. The deadly nightshade of that family deserves its name—among other bad solanaceous stuff. Commercial silkworms must have their mulberry leaves, but other silkworms prefer other plants. And so on.

We humans have developed considerable fondness for some of these herbivore deterrents: plant secondary metabolites. They include the tannins that give tea its astringency and a large fraction of the components of every spice shelf. But many more are as bad for us as they are for other herbivores. Among these are some of our famous poisons and near poisons:

compounds such as curare, belladonna, and digitalis, as well as most of the obscure poisons invoked by the authors of murder mysteries.

The second tactic for reducing herbivory, physical deterrence, and the third, low yield, blur together more than either one does with the first (chemical defense). An herbivore can't evade a cost-benefit game. It wants the maximum yield in nutritional terms for the minimum cost in its time and energy. At the very least, energy yield must exceed energy cost—an issue we face when deciding whether to use corn-based ethanol as a fuel. So the game for the leaf consists of being hard to obtain and process and then, as near as is practical, to constitute pure junk food.

"Hard to obtain" has several dimensions. Ground-based herbivores who have to reach their greens can be foiled by plants that supply little or no low-hanging food. Among mammals, obligatory leaf herbivores tend to be fairly large, and relatively few of them can do much climbing. But the arms race remains evident. Why else would giraffes have such long necks, necks long enough to restrict their maximum speeds and loco-motory agility, necks long enough to require major adjustments of their hearts and blood vessels? Less commonly recognized, giraffes also have remarkably long tongues that they can extend further upward by tip-ping their heads back, the better to wrap around still higher leaves. By the same token, why might prickly-pear cacti (*Opuntia echios*) in the Galápa-gos grow much taller and sprout their lowest branches at greater heights on islands where land tortoises are present?

Another approach to reducing herbivory is to maximize masticatory messiness. Cellulose itself is tough stuff, and the lignins that lace the var-ious cellulosic compounds together help by increasing the sizes of tooth challenging structures. For our own palates, we select particularly non-fibrous varieties of minimally fibrous leaves. Even so, only the thinness of leaves (lettuce, etc.) or our willingness to swallow fairly large chunks of fibrous stuff (celery) keeps us from equating plant material with the toughest cuts of meat. We recognize some vegetables (root vegetables, especially) that haven't been harvested early enough by the extra work it takes to chew them up—they contain too much woody cellulose and lig-nin. But woodiness may be the least of the problems.

Many plants lace their leaves with silica, the very essence of grittiness. Just as cellulose is the earth's commonest organic compound, silica in its various forms is its commonest inorganic. Its formula could not be simpler: SiO_2, silicon dioxide, analogous to CO_2, carbon dioxide. We know it as quartz, sand, glass. It's hard stuff, with a Mohs' hardness value of 7 on a scale that runs from 1, talc, to 10, diamond.*

Of negligible toxicity in any chemical sense, silica puts grit in the gears. Eating plant material with more than a little silica simply grinds down the teeth of herbivores. Our hardest corporeal material is the enamel of our teeth, with a Mohs value of 5. So eating herbage scratches and slowly wears down teeth. Anthropologists judge what our ancestors ate in part by observing the scratches on their teeth, and our toothpastes (as well as our scouring powders) often contain silica as an abrasive cleaner. The worst offenders are the grasses, which contain several percent silica. More work has been done on the difficulty of grazing on them than on the leaves of trees. But silica occurs in other leaves as well—grasses just make fewer chemical deterrents to confuse matters.

Grass-eating herbivores such as cows and horses have teeth that put materials of varying hardness side by side—enamel and dentine mainly, as in figure 13.7. That produces teeth that don't smooth off as they wear. Each material wears down, but the harder one persists in protruding above the softer one, the way sanding a piece of oak with its heavy grain never makes it quite smooth. Some herbivorous mammals make teeth that grow continuously. Elephant jaws generate teeth in the back, the teeth move forward over time, and, well worn, they're reabsorbed in the front—if you see an elephant skeleton in a museum, look for the increase in tooth wear from back to front. Curiously, even the longest-lived vertebrates never make the hardest compounds that occur among animals. Some limpets (a group of shelled but uncoiled snails) rasp algae with teeth of silica itself, and chitons (also mollusks) do so with magnetite, an oxide of iron that has a hardness of 7.5.[9] Some vertebrates use magnetite as a magnetic sensor for orientation—but not for teeth.

---

*Mohs' scale gives the results of scratch tests: a material with a higher value will scratch a lower-value material. It's an old measure, created by Friedrich Mohs in 1812, and peculiar among our various measures in being an ordinal scale. So one can speak of "harder than" or "less hard than" only, as it puts materials in order but says nothing about the size of the intervals between the values.

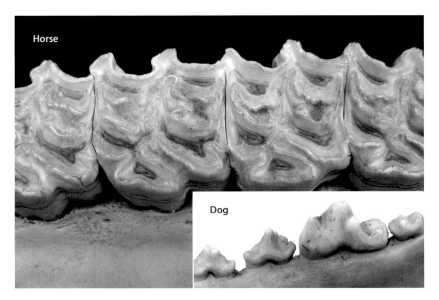

FIGURE 13.7. The peculiar, wear-tolerant grinding molar teeth of a large herbivore, in this case a horse. A dog (*inset*) has quite different molars. Ours are grinders, but without the horse's extreme stratification of materials.

Again, the big players in the herbivory game are larval and nymphal insects. Adults do less immediate damage, in part because adults don't grow—no flying insect, meaning no fully winged adult, ever molts again. In addition, and more about this to come, leaves don't provide an effective energy source for flight. About the only leaf-destroying adults that come to mind (my mind, anyway) are the leaf-cutting ants of tropical forests. They snip off pieces that they carry back to their nests as fodder for their fungal farms (fig. 13.8), what we'd call animal husbandry (but more literally wifery, since the workers are females). Ants, of course, aren't frequent fliers.

Insect mandibles aren't all that hard, but then they're short-lived structures, replaced in each molt. Nonetheless, they're about the hardest parts of an insect's cuticle, with their basic material, chitin, augmented by various compounds of zinc and manganese. Again, one wonders why vertebrates never use such hard, heavy-metal compounds. Our metabolic machinery routinely deals in heavy metals—quite a few are obligatory dietary components, critical for synthesizing certain enzymes.

FIGURE 13.8. Leaf-cutting ants heading home on one of their roads, the latter marked by my blue dotted line. Outbound ants don't show up in the picture, but they were present as well. The photograph was taken at the Smithsonian lab on Barro Colorado Island in Gatun Lake, Panama.

When the Going Gets Tough

Hardness as measured on Mohs' scale addresses only one of the mechanical problems of an animal that bites, shears, or masticates leaves. Besides causing wear on the cutter and grinder teeth, the design of leaves ensures that eating them takes a lot of work. What nature has joined, let no tooth put asunder.

Here we need another quantity, in essence a measure of how much work it takes to make new surface. The quantity commonly goes by the names toughness, fracture toughness, or work of fracture. It's not the same as stiffness, the resistance of a material to deformation, nor to strength, the greatest stress a material can take before breaking. Confusingly, in some accounts toughness represents the energy something can absorb before failing, such as the energy that might be stored for the shot when a bow is fully drawn, not the energy absorbed as the thing actually fails. Ambiguity can be avoided by using the term *fracture toughness* for what's meant here. Again, it's the work needed or energy input required to make new surface as something parts company with part of itself.*

Few of us notice, but making fractures that create only a small amount of new surface takes less energy than making ones producing a lot of surface. A glass rod breaks easily and cleanly, at least once scratched. A fiberglass rod, made almost entirely of the same material, breaks with great reluctance and jaggedness, whether or not it's initially scratched. Materials such as aluminum foil tear cleanly and therefore—we can now assert—easily. Tearing fabric takes more work and leaves a fuzzy edge that can't be confused with a cut edge—fibers protrude after we've torn them or undone their interdigitations. Yes, leaves tear, but more like fabric than like aluminum foil. Maybe they take a little less energy than fabric does, but they're wonderfully resistant to tearing in a straight—and thus short—line.

We've now entered yet another area of physical science, a branch of engineering known as fracture mechanics. For all its obvious practical importance, fracture mechanics remained puzzling well into the twentieth century. All too often, materials and structures fractured with far less provocation than ordinary mechanical measurements of material strength or calculations from chemical bond strengths led us to expect. Teeth popped off gears, and metal ships broke amidships under all-too-modest loads.[10]

The particular subject, brittle fracture, asks why in some circumstances cracks extend much more easily than in others. (Metals actually resist brittle fracture better than other hard materials such as glass.) The do-it-yourself looks at one mechanism by which failure-by-fracture can be, if not avoided, at least postponed to higher levels of stress. The mecha-

*Looking at the dimensions or units provides an easy way to tell which version of toughness is meant. Force per area (as newtons per square meter or pascals) or work per volume indicates work or energy absorbed up to the breaking point. Force per distance or work per area indicates work or energy absorbed as something breaks, the present issue. The units are the same as those of surface tension, unsurprisingly inasmuch as surface tension opposes the creation of new surface, just as does fracture toughness.

> **Do it yourself:** Getting a feel for the centerpiece of fracture mechanics could not be easier. Tear off a piece of aluminum foil, a nicely homogeneous material. Fold it in half to produce a virgin edge, then fold, and then pull on the ends of the fold, noting its resistance to tearing. Next, with a pair of scissors make a nick no more than about half an inch (1 centimeter) in the middle of the fold. Now pull on the ends of the fold—the foil will tear in half almost effortlessly. Obvious, sure, but less so is what happens when you remove more foil (on another piece) by punching a round hole with a paper punch at the end of a similar nick. Same material, but it now takes more force before tearing. The round hole at the bottom of the nick has mysteriously restored much of the original tear resistance.

nism depends on the way a crack extends at its tip. The sharper the tip, the more concentrated the energy available to do the job.* Spread that energy around, as does a rounded end on the crack, and it less effectively instigates extension of the crack. Punching a hole, removing material, can make a structure stronger in this eminently practical sense if the hole rounds the sharp end of a crack. Put simply, a round hole can keep a crack from extending.

At this point, it's uncertain whether leaves make themselves harder to rip apart with this crack-stopping mechanism. My casual survey of the leaves where I live suggests that maybe some do and some don't, or maybe their designs don't reflect this particular problem at all. Leaves with shallowly saw-toothed (serrate) edges, such as those of beeches (figure 13.5 again), may have sharp tips on the edges, but they often have valleys rather than sharp canyons between them—as we now expect. Oaks include trees

*The relationship between the sharpness of the crack tip and the amount by which the force at the tip is concentrated by the crack is quite simple, at least for brittle materials—that is, those that don't stretch under load. It's expressed in terms of the stress concentration factor, C_{stress}, which varies with the length of a crack extending from the surface, l, divided by its tip radius, r:

$$C_{stress} = \sqrt{\frac{l}{r}}.$$

with particularly lobed leaves, but the indentations (sinuses) between the lobes are inevitably rounded—again as we now expect. While some of the maples and our very common sweetgum also have fairly sharp corners beneath their indentations, they may play another game. They might reinforce the bottoms of those indentations. But instead of that, they seem to do something a little more subtle: they make a little extra surface at these locations so, although still flat, they're no longer two-dimensionally planar.

A leafwide version of the device appears to be widespread. Leaves may be flat, but their surfaces don't usually fall on a proper plane however that plane, like a sheet of paper, might be curled, creased, or crumpled. (So no trick of origami can fully capture the form of the surface.) First, the baseline. In ordinary Euclidean geometry, a pair of parallel straight lines on a surface remains parallel however far they're extended. By contrast, leaves may have extra peripheral surface, giving them what mathematicians call negative curvature and describe as saddle shaped. In that geometry, two initially parallel straight lines diverge when extended in either direction. Or they may have positive curvature, as does the surface of a sphere. Here two initially parallel straight lines will converge and eventually cross, as do the lines of longitude as they extend away from the equator of the earth. Figure 13.9* shows a particularly clear comparative pair.

My casual survey suggests that a leaf can even combine the two kinds of non-Euclidean surfaces. The part nearer the attachment of the petiole, the inner part, may be positively curved. Parts either out near the far end of the leaf or near the bases of even broadly rounded indentations often go in for negative curvature. But I cheerfully admit to the chance that I see what I'm looking for as much as what's objectively present. I may just see what my functional explanation, the following, needs if it's to work.

Think about the advantages of each of the deviations from planarity. Reduced edge—the curling of positive curvature—can give greater resistance to bending with negligible extra investment in material. That

*For the botanically savvy, note that negative and positive curvatures do not just redescribe what have long been called revolute and convolute leaves. These latter terms refer to lengthwise curling, backward (or downward) and frontward (or upward), respectively. Perfectly Euclidean surfaces, such as the surfaces of cylinders, can do that kind of curling.

FIGURE 13.9. Two non-Euclidean leaves. The one on the left, *Thunbergia mysorensis* (from the greenhouse at Duke University), has extra edge and thus negative curvature. The one on the right, an ornamental holly, *Ilex burfordii*, has extra middle and thus positive curvature.

> **Do it yourself:** A quick test could not be simpler. Simply press a leaf between a sheet of glass or stiff transparent plastic and an underlying flat board. Negative curvature reveals itself as radial wrinkles or folds that run out to the very edges of the leaf. Positive curvature produces wrinkles (rarely folds) on either side of the midrib that don't extend all the way outward to the edges. These wrinkles come from the way a relative shortage of leaf edge pushes the left and right outward edges inward, toward the midrib.

should translate into less likelihood of drooping downward under self-loading. Bending will be more of a problem near the base of a leaf, from which more leaf extends outward (greater force) and extends outward a greater distance (greater moment arm). Extra edge—negative curvature—should make tears harder to initiate, as might (and on occasion does) happen when a strong wind strikes a leaf. With negatively curved edges (or vulnerable parts of edges), tensile forces won't concentrate to the same extent.

Extra edge on photosynthetic organs seems to have been investigated only in the blades of macroalgae that have undulate edges.[11] For them, ruffling appears to solve problems such as flow-induced clumping that shouldn't afflict the broad leaves of trees.

Maybe natural selection balances the relative advantages of increased effective stiffness of the blade achieved with positive curvature against the extra resistance to tear initiation gained from negative curvature. Or perhaps a leaf might use both. Stiffness should help more near the base, while resistance to wind-induced tears will be more useful at the indentations between lobes due to local force concentration. And tear resistance will matter most near the far end—for the same reason that flags flutter and fray at their far edges.

Leaves do rip up in severe storms, especially early in the leafy season. Perhaps the lack of any obviously specific pattern of tearing tells us something about how their designs equalize the areawise chance that a tear will initiate. This notion of balanced design, by the way, amounts to asserting that natural selection will produce the equivalent of Oliver Wendell Holmes's deacon's wonderful shay:

> Have you heard of the wonderful one-hoss shay,
> That was built in such a wonderful way
> It ran a hundred years to a day
>
> . . . .
>
> You see, of course, if you're not a dunce,
> How it went to pieces all at once,—
> All at once, and nothing first,—
> Just as bubbles do when they burst.*

C. Richard Taylor and Ewald Weibel coined the term *symmorphosis* for the concept. (That's a good search term, although I don't want to revisit the controversy it initiated.)

Another good way to minimize crack propagation comes down to an evasion of the basic assumption of brittle fracture, that of brittleness. Nick and then try to tear a piece of Saran or other plastic wrap. It resists tearing by stretching, in essence by not being brittle. In that way it lets the load do the work of rounding the tips of any nicks or other indentations. Leaves probably take little if any advantage of this latter scheme, although some macroalgal fronds appear to do so.

*From "The Deacon's Masterpiece," the first three lines and four near the end.

Leaves have yet another way to limit tearing. Cracks or rips don't like to pass through discontinuities in the mechanical character of a structure. Look at a cracked sidewalk or pavement, or else at figure 13.10—three-point junctions could not be more common, while finding four-point crack junctions takes careful looking.

FIGURE 13.10. When pavement or other brittle material cracks, three-crack junctions far outnumber four-crack junctions.

What's happened? A crack met another crack that crossed its potential path, and it found the other crack enough of a barrier to stop its extension. A crack will act as a crack-stopper. The veins on leaves represent just such abrupt mechanical changes, and a rip is likely to stop advancing when it meets a vein. Rip a leaf, and the result zigs and zags, changing direction where the extending tear hit veins. If veins block direct advances, then moving more crosswise to the tearing force becomes easier than moving with it. The consequent increase in tear length, those zigs and zags, means more work has been done making surface.

At this crack-stopping game, the champions must be grasses. Tearing a leaf of any grass, even of a bamboo, crosswise is remarkably difficult. Nick a grass leaf, and it does lose tensile strength—but just in proportion to the depth of your indentation.[12] So an herbivore can't rip grass leaves as it eats. Instead, it has to slice them, chew its way across, or else pull the plant out by the roots. Different grazers use different tactics with different consequences. In particular, the root pullers make a mess of the land, exposing it to increased erosion, possible desertification, and so forth. Slicing and chewing are more benign. Grasses, unusually, grow from their bases and not their apices—which is why lawns tolerate repeated mowing. Not only are slicing and chewing less injurious to grass leaves, they may be a practical necessity, since many ungrazed grassy habitats slowly shift from range to forest.

And the Payoff Is Small

That about says it. After a leaf is munched, crunched, and found to be nontoxic, the third and final line of defense comes into play. Ordinary leaves simply don't provide much in the way of nutrition. Not that they truly resemble our junk food—nothing but calories. In a sense, they're just the opposite: they may provide some protein and other useful stuff, but they yield precious little energy relative to their mass, not to mention relative to all the work they make herbivores do to ingest them.

The worst part must surely be all that cellulose. Its malevolence goes well beyond the difficulty of shearing and masticating the cellulose-lignin composite material. So frustrating—cellulose as a molecule looks almost exactly like starch, the very bread of life, and glycogen, our own storage form of carbohydrate. Almost—but the bonds between the individual sugar units have a different orientation (beta-glycosidic rather than alpha-glycosidic, meaningless terms to the nonbiochemist). Somehow that minor difference renders the cellulose indigestible by nearly every animal. Only by harboring internal microorganisms and outsourcing (insourcing?) the task can an animal get any nutritional value at all from eating cellulose. And the microorganisms, no surprise, extract their processing fees as unapologetically as do bill collectors and stockbrokers.

Famous wood eaters such as shipworms (bivalve mollusks, really) and termites can't digest wood themselves, nor can great leaf eaters such as garden snails and ruminant mammals handle cellulose unassisted. It's possible that silverfish (wingless insects, abhorred by all librarians) make a proper enzyme, a cellulase, to do the job; but if so, they're the rare exception that makes a peculiar situation even more peculiar.[13] Still more peculiarly, a few animals, ascidians (tunicates) in particular, synthesize cellulose—they can make it but not break it. For us, it's the main element in the "insoluble fiber" noted on the labels of packaged foods, entirely indigestible but at least good stuff for our intestines. For reporting energy (calorific) yield, the US Department of Agriculture quite properly gives insoluble fiber a value of zero, rather than the 4 kilocalories per gram yielded by normal carbohydrates. The latter is the value we get if measuring the total oxidizable material in some food, but only when burning wood for fuel do we realize that energy.

Even the cellulose digesters, the creatures with symbiotic microorganisms, pay at several levels for handling the material. Those for whom leaves provide the main source of energy are a dull lot that spend their lives eating, and eating, and eating. I think no animal that lives exclusively on ordinary leaves manages to fly—processing them is just too costly in machinery and time for life in that fastest of fast lanes. Yes, some caterpillars live on leaves, but the resulting adult insects do not. In their metamorphosis they've reabsorbed and reinvested the extra digestive equipment. Some flying birds eat leaves, but they don't do so exclusively. This answers the question implicit in a scatological rhyme I recall from my childhood (and recently taught my grandsons):

Birdy, birdy, in the sky,
Why you poo-poo in my eye?
I'm a big kid, I don't cry,
But I'm just glad that cows don't fly.

Incidentally, all the leaf eaters not only have disproportionately large digestive systems but also discharge a disproportionately large volume of feces. Mucking out the barn is no trivial task, and before the ubiquity of internal combustion engines, cities coped with vast volumes of horse manure—plus numerous organisms that found it to their liking.

Not all that many mammals live exclusively (or nearly so) on leaves. A large number consume some leaves, but the vegetarians among them (most humans are essentially vegetarians these days) also eat the more nutritious parts of plants: buds, roots, seeds, fruits, and so forth. Fully folivorous mammals tend to be medium- or large-sized animals, presumably because the diet is impractical for small warm-blooded creatures, owing to the high cost of their thermoregulation. About the smallest is the marsupial quokka, in the same family as kangaroos, which weighs 5 to 10 pounds (2 to 5 kg)—roughly cat sized.[14] No small rodents need try.

Not only are the digestive systems of folivores (and predominant folivores) large, they incorporate lots of specialized devices to provide happy hostelries for all those co-opted bacteria, fungi, and protozoa. Ruminants have large, multichambered stomachs, horses have large outpocketings of their intestines, and kangaroos have convergently evolved somewhat ruminant-like stomachs. Even rodents and rabbits reingest some of their feces to maintain their digestive flora. And so on—ordinary leaves make lousy food, even for those animals that house cellulose composters. And for those who don't—eat your greens if you want to lose weight but still feel full when getting up from the table.

Making low-energy leaves must give plants two separate benefits. Discouraging herbivory may actually be the lesser of these advantages. If leaves are produced and then discarded annually, making them on the cheap should be especially important. So plants should economize on leaf structure, even beyond nature's usual preference for economical designs. In addition, they should immediately export the products of photosynthesis both to discourage herbivory and to cheapen these throwaway structures. Reducing live leaf weight might confer a third benefit, if a lesser one: reduction in the aboveground weight that has to be supported.

Winding It Up

14

THE SHADE OF A FOREST feels different from the shade of an awning. The shade of pines isn't quite like the shade of oaks, and neither resembles the dappled, shimmering shade of cottonwood or aspen. Forests and groves have their distinctive smells as well. Some of this book was written amid the Douglas firs of the Pacific Northwest; the rest happened beneath the loblolly pines surrounding my house in North Carolina—each has its special odor, and neither could be mistaken for a hardwood stand. The "smokiness" of the Great Smoky Mountains comes not from any haze of fire but from a mist of odoriferous particles produced by its trees.

Even forest floors have their individuality, not just in appearance but in what must have mattered to Indian moccasins, what one might call "foot-feel." Even the sound of wind varies from forest to forest and season to season. To me, each provides its peculiar pleasure, a greater one—a matter of personal history and taste—than I enjoy at any art museum. Having some sense of the scientific, asking the kinds of questions posed in this book, wondering about both the deep and the recent history of a forest—none reduce the pleasure that I first enjoyed as a youngster, hanging out in a swampy patch of mixed hardwood forest in the Hudson River valley.

As writer, then, I have no fear, and as reader you need not worry, that this unrelievedly scientific account might detract from any aesthetic experience. It might even enhance it, but that depends on your particular turn of mind. In the end, which this chapter marks, the book has been about science and nothing but science, one long argument (Darwin's phrase, incidentally, in his *Origin of Species*) that biological imperatives cannot transcend physical reality, that biological success comes both from heeding the constraints of the physical world and from capitalizing on its opportunities. In the end, all the specifics that filled this account give us a perch from which to look at some larger concerns, linking the biological with the physical and finally touching on the social.

Size, Scale, and All Those Physical Factors

Neither a big creature nor a big bridge will do well if it's simply a scaled-up version of a small one. Any design bumps up against a size limit, as we noticed for the heights of trees. Diffusion works wonderfully well for microscopic systems but takes nearly forever on any scale much larger. So diffusion suffices to move material within animal cells, but plant cells (which are larger than animal cells), trees, and we ourselves must add pumping to our internal transport systems. The Reynolds number includes a factor (two factors, if we include speed) for size, telling us that large-scale flows of fluids tend to be turbulent, while small-scale flows are most often laminar. So sap and blood flow laminarly in the conduits of plants and in your blood vessels, but almost all the flows in your household plumbing are turbulent.

For nearly all practical purposes, the physical factors that might be relevant for just about anything—be it living or nonliving—depend on its size. And organisms span a huge size range, with the largest roughly a hundred million times longer than the smallest—from bacterium to whale, coincidentally of about the same shape. Size and the parent subject, scaling, have been a quiet subtext in every chapter here. It now needs to come out of the underbrush.

Most pervasively, the interrelationships of length, surface area, and volume depend on size, unless shape changes concurrently, systematically, and drastically with size. A sphere ten times the diameter of another has a hundred times as much surface and a thousand times more volume. The same holds for a cube or a cylinder that's, say, twice as long as it is thick. Put more formally, areas of objects of the same shape increase in proportion to their lengths squared; their volumes, to lengths cubed.*

*The usual way to express these relationships uses a convenient if uncommon symbol, $\propto$, read as "is proportional to." It takes the place of a constant and an equal sign where the constant isn't immediately relevant. Since the proportionality factor for scaling relationships is an exponent, it appears as the slope on a graph of the two variables in logarithmic form. For instance, using length, l, surface, S, and volume, V,

$S \propto l^2$ so $\log S = 2 \log l + C$

$V \propto l^3$ so $\log V = 3 \log l + C'$.

$V \propto S^{1.5}$ so $\log V = 1.5 \log S + C''$.

For determining the actual exponent of proportionality from real data, you plot the logarithmic versions and measure or compute the slope of the best line. For biological

A bit more subtly, the larger the object, the more volume it will have *relative to its surface area.* A sphere ten times the diameter of another, the one above, has a volume-to-surface ratio ten times greater, a surface-to-volume ratio (put the more usual way) a tenth as much. The whale above has a hundred million times *less* surface, relative to its volume, than does the bacterium. In essence, the whale has a lot of inside but relatively little outside, while the bacterium has very little inside but a relatively huge outside.

This size-dependent surface-to-volume ratio must bear on the question of how high a tree finds it cost-effective to grow. The tree absorbs light as a surface, the area that its crown presents to sun and sky. At the same time, its structure constitutes a volume—comprising roots, trunk, branches, twigs, and leaves. If everything simply enlarges as a small tree becomes a big tree, both its overall surface-to-volume ratio and those of its components will decrease—all surfaces, all volumes. The big tree will be able to absorb less light relative to the volume of tree that has to be maintained. Thus as the tree grows, the resources available for further growth and, most important, for reproduction will gradually diminish— again relative to the volume present.

As ought to be obvious by this point, the way a real tree changes the sizes of its parts as it grows can be described by no simple formula. The volume of trunk and branches needed to support a given area of crown goes up even more drastically than a surface-to-volume rule would pre-

matters, we most often plot body mass versus whatever else is of interest. Units don't matter much; they affect only the constant that has been evaded by using a proportionality instead of an equation. Radius and diameter give the same slope, as do mass and weight. Again, and crucially, the relationships in the formulas above assume no systematic size-dependent shape change.

dict. Height imposes a special cost of its own. Winds become stronger with distance from the ground, and the drag they impose acts with longer lever arms. Branches become longer, and their added weight acts further outward from the trunk to which these cantilevers attach. Water has to be raised to supply leaves. Water is dense and therefore heavy, and trees, particularly broad-leafed ones, raise a lot of it from their roots to their crowns. In addition, the bigger tree needs a wider area of water- and nutrient-absorbing roots and either a wider root plate or a more substantial taproot to avoid toppling in storms. It takes little more than a casual glance to recognize that trees don't grow everything in proportion, maintaining their shapes as they get taller until they tickle some limit set by diminishing returns. It's no trick at all to tell a sapling from a mature tree in a photograph that has no indication of its scale, such as a ruler or other object of recognizable size. Trees know, so to speak, about scaling, and they make adjustments, lots of them, as they enlarge.

Trouble is, they also know about their individual situations, adjusting to accommodate their personal particulars and limitations and thus deviating further from strict scaling rules.[1] As it grows taller, a tree ought to get fatter, whatever the specific scaling rule. But palms and bamboos entirely lack the machinery for radial growth—to take an extreme case.

Recognizing the general problem—the conflict between geometry and biological function—and the necessary general solution—size-dependent shape change—helps us to recognize in turn a tree's specific responses. More specifically, we look for change that can't be attributed to accidents of ancestry or constraints imposed by growth processes per se. That's an important approach for the biologist, our version, perhaps, of the old philosophical principle known as Occam's razor. One's mental toolbox holds no more useful implement: try the simplest, most straightforward explanation first. Not only will it usually be the easiest, but it's the one most likely to be correct. Our version of Occam might be put in reductionist terms: rule out geometric, then physical, and then chemical explanations before invoking biologically specific ones. Of course, I cheerfully admit the influence of personal history. The first bit of science I pulled off consisted of showing that a phenomenon (admittedly not a big-deal phenomenon) resulted from the direct action of fluid mechanics on a biological system rather than some special neuromuscular reflex adjustment, as had been assumed.

Leaves don't scale the way most other biological systems do—they retain even less trace of proportionate increase in length, width, and thickness. Big trees don't have bigger leaves than do small ones. Instead, they have *more* leaves—just as large animals have more cells, not larger cells, than do small ones. Beyond that, broader leaves aren't also thicker ones. If anything, smaller leaves tend to be thicker, whether we're looking at the sun-shade differences among oak leaves or more generally at the broad leaves of both ordinary forests and deserts and semiarid areas. Even ignoring the thick leaves of dry habitats, we find that while surface area may scale with something close to the expected length squared, volume comes nowhere near scaling with length cubed. By this point, you ought to have no difficulty thinking up a plausible physical explanation for the peculiar way leaves deviate from the basic geometric relationship between surface area and volume.

In terms of scaling, leaves have another peculiarity: is an ordinary leaf big or small? The trivial question takes on a wholly different meaning when asked of leaves than when asked of mice, monkeys, and mastodons. A leaf might be two hundred times wider than it is thick. That's a drastic enough contrast so that different physical rules for a leaf's directional processes must apply in their different directions. Both diffusion and conductive heat transfer work fairly well between the leaf's top and bottom surfaces. Diffusion receives a little help from having some gas rather than only solids and liquids within the leaf; conduction receives a little hindrance from that same difference. But for transport between top and bottom, neither makes a great deal of difference in this broad-brush view. With the aid of intracellular circulation, cyclosis, an ordinary leaf is thin enough to have no major crosswise transport or heat transfer problem. When I measure the temperatures of the top and bottom of a leaf, they differ at most by about a Celsius degree.

On the other hand, the internal structure of leaves, never mind the physics involved, tells us with no ambiguity that diffusion can't alone do the job of moving material lengthwise and breadthwise. Veins ramify throughout, veins equipped with conduits through which liquids flow. We're three-dimensional creatures, as are our circulatory systems. By contrast, leaves are two-dimensional structures, at least with respect to nonmicroscopic processes. Their equivalent of our circulatory system branches in two, not three dimensions. They don't move enough fluid through their veins to augment lengthwise and breadthwise conductive heat transfer the way we do. So they're stuck withstanding striking differences between center and edge temperatures, worse ones, relatively, than those between our cores and fingertips.

You should at this point be impressed by how many physical factors bear on the life and effective design of both individual leaves and whole trees. At least above the molecular level, these factors themselves don't depend on size. Diffusion coefficients remain diffusion coefficients, radiant emissivities remain as they are, the same goes for the bulk modulus and thermal expansion coefficient of air, environmental temperatures, gravitational acceleration, and on and on. Conversely, I can't think of any one of these whose effects don't, either directly or indirectly, depend on the size of the leaf—or the size of any other biological entity. So the system that enlarges without change of shape rather than one that does change should be recognized as something rare and exceptional, something that needs some special explanation.

Nor is our species any different. Compare a large with a small mammal. The large one has, relatively speaking, somewhat more skeleton, about the same muscle mass, slightly less liver and kidney, and a lot less brain. Its voice is deeper, and its heart beats less rapidly. Although it consumes relatively less food, it runs or swims faster. For each of these features, physiologists now understand fairly well the interrelationships between the way it scales, the function it serves, and the operative physical imperatives.

The Functions of Features

More often than not, a structural feature of an organism serves multiple functions. Not only do our hearts pump oxygen-distributing arterial blood to our entire bodies (including themselves), they also provide the hydraulic motors to operate the initial filtration stage of our kidneys and to initiate erection of penises, and they transfer information in the form of hormone levels. A leaf, unusually, has one transcendent function: photosynthesis. But looked at more closely, that atypicality disappears, the way a note in a symphony or a brushstroke in a painting retains nothing of the greatness of the whole. So at finer levels the structures of leaves may be as multifunctional as those of any other biological system.

I myself find little of interest in biological structures per se. They're only slightly grabbier than (for me, again) the anatomical terminology with which we describe them. Function contributes to reproductive success; structure does only as it serves function. That's why the present story has focused on function.

Which brings up another question-in-conclusion: what does a feature do for the organism? That's asked by children, by biomimicry-centered design courses, and by those of us afflicted with insatiable curiosity. Such a question holds surprising subtleties and complications. In a sense, it's both central and unique to biology, one that a person can't ask in the same way about a rock or an ocean, and one to which the answer is far simpler for the function of the items we humans design. Biological features have evolved, and that carries an intrinsic presumption of functional significance. But if a feature serves multiple functions, then its design must involve compromises. So part of our job consists of figuring out the relative importance of the relevant functions, the constraints each imposes on the way something is designed to achieve the others.

Of course, a feature may do nothing at all. I offer the quaking of the leaves of aspens and other poplars in light breezes as a possible example. What catches the eye may be purely incidental to something else that matters adaptationally, that is, reproductively in the broadest sense. Quaking may just be an instability in low winds incidental to good drag-reducing reconfiguration in strong winds. The late Steve Gould was fond of pointing out that nothing mandated function, that the assumption of functional significance carried serious hazards. I'd not go quite as far as he did, but at some level and to some extent, he has to have been correct.

Or a feature might be bifunctional or multifunctional in ways that make it difficult to sort through as we ask which function—or functions—most directly reflect natural selection, that is, which are evolutionarily primary. The conduits, xylem, that carry water up vascular plants of all sizes (why we call the pines, palms, poplars, and potatoes vascular) also make up the structural material we know as wood. Which function is the more important in determining the size and details of xylem-wood? The answer to that and similar questions may be far from obvious and far from easy to determine by direct observation, experimental manipulation, or computational simulation.

Or a feature may be multifunctional in a more limited sense in that one function defines its necessity—like a photosynthetic leaf—but the feature has to do much more to deal with the disabilities, the downsides, of carrying out that function. Leaves run the risk of getting too hot, too cold, too draggy, and so forth. Secondary functions can be fully as structure-determining as the primary one. Because I wrote a book on fluid flow in biology, I'm asked once in a while about how the design of some organism minimizes its drag. Except perhaps for a slowly descending seed of a dandelion or milkweed, for a diving falcon, or for rapidly swimming aquatic animals, drag minimization won't be the central function determining design. So I have to point out that drag may or may not matter, or at least not matter enough to compromise more crucial activities. Trees would have vastly less drag if they lacked leaves! I've tried to bear in mind this caution, since I've usually studied one or another of these secondary functions, the necessary evils that tag along with the truly critical aspects of life.

Leaves and More Leaves

Leaves, even those of tall trees alone, come in a bewildering diversity of forms. So what? To the biologist, that diversity itself carries significance. We might guess that diversity has increased in fits and starts ever since life first evolved on earth. That seemingly obvious trend seems less self-evident after a little thought. Evolution keeps selecting for what works well and rejecting what works less well, so its products should get ever better, ever closer to optima. If so, then maybe evolutionary designs should become fewer over time. After all, optimal designs should be less numerous than suboptimal designs. Put another way, the evolutionary process should be more efficient in rejecting than in creating. So why do leaves still come in so many forms, even among adjacent trees presumably subjected to the same vicissitudes of climate, herbivory, and mutual competition?

The question looks no less puzzling in its sharp contrast with the way human technologies progress. We innovate, then we gradually shake down and standardize on a few designs, ones either intrinsically superior or with sheer market dominance. In 1920, some automobiles had internal combustion engines, some had external combustion (steam) engines, and some had battery-powered engines, ones in which the original energy source wasn't on board. Now we wonder about the practicality of any shift away from the universality of internal combustion. (Non-plug-in hybrids depend as much on internal combustion as do ordinary cars.) In 1960, large commercial airplanes used jet, turboprop, and piston engines; now jet engines power all of them. The gauges of the tracks of railroads, the sizes of film for nondigital cameras, the operating systems for computers—all rapidly diversify and then get pruned back to one or a few systems. Can we draw any analogy with evolutionary change in nature? Historians of technology and paleontologists give them equivalent names: lock-in and the privilege of incumbency. I prefer to remain agnostic, but you can make your own judgment about whether the processes are analogous.

Some millions of species, each looking out for its own interests as natural selection dictates, generate a multidimensionality that's either dazzling or mind-numbing, depending on one's attitude. Physicists often disparage our continued inability to come up with the biological equivalents of the Newtonian laws. (Their own occasional efforts to help us haven't been especially productive.) Too few of us biologists have the self-confidence to counter with the suggestion that physicists have had a history of putting aside areas that weren't satisfyingly law-abiding. They commonly left messy stuff such as fluid dynamics to the engineers, practical people less often in a position to pick and choose. But their successes achieved through merciless pruning don't imply that we should sweep the multidimensionality of biological determinants under the leaf litter.

Maybe we shouldn't even be surprised by all that biological diversity and complexity. A good argument can be made that the more factors that bear on whatever constitutes good design, the more optima should exist. These might not be absolute optima, but we should at least expect a landscape of nearly equivalent suboptimal peaks. If force and mass are all that matters, we can assert a rule (Newton's second law) linking them with nothing more than the value of acceleration. If all those factors I've been carrying on about in this book matter, then the optimal form of something such as a leaf should not be so easily and uniquely specified.

Messages of the Messiness

Ignoring this messiness would bowdlerize the world of our leaf, because all that diversity and complexity carry general guidance for the perplexed biologist. First, though, a word of caution (or skepticism) about an all-too-common contemporary ex-cathedra pronouncement. In whatever wording it comes in, it asserts that nature has tried everything. If it doesn't exist in nature, then it's not a good thing—or, since we're talking about revelation, maybe "A Good Thing." The counterargument? You can easily think of forms that leaves might assume but never do. How about multiple petioles, the way we use multiple girders to support some cantilevered balcony? How about more radical trussing systems than just folds or veins embedded in a lamina? How about a photosynthetic system that absorbs and uses rather than rejects green light? The same goes for entire trees. Why not trees that reabsorb and reinvest the wood in the cores of their trunks—beams and columns make the most of material when fat and hollow, not thin and solid. Or treelike organisms that get basic support from something other than wood, perhaps proteins stiffened with some calcium salts like our bones, or metals like those of our larger constructions? And that doesn't touch the larger question of whether evolution could possibly test all possibilities, even with infinite time to do so.

No, what we know of the evolutionary process simply does not support an argument for global optima in nature's designs. Evolution, as introduced in the first chapter, wanders, misses chances, reinvents wheels, has trouble making radical alterations in its designs, and so forth. Until modern humans came along, with our large-scale, culturally-transmitted technology, evolution had the game to itself, the ultimate winning situation. It may be a lousy scheme, but it does have persistence. We confidently expect it to go on long after we've blundered into some black hole of extinction—most likely one of our own making.

A more defensible message emerges from the sheer creativity of the evolutionary process, however constrained it may be. It has generated a system that's extravagantly diverse from just five nucleic acids, twenty-odd amino acids, a few basic polymers, one primary photosynthetic chemical, no metals, and lots of other limits to its basic components. If random exploration can do such impressive things, then, unless it self-destructs, our own technology might go far beyond its present level of sophistication. After all, it has far fewer intrinsic limits, and it has the anticipatory power of rational science.

Another message comes from the quantitative limitations we can recognize within all that diversity, especially all those size-dependent factors that have wandered through the present account. Here basic evolutionary creativity plays a minor role, so we needn't worry much about hypothetical evolutionary jumps or difficulties in growing some hypothetical form from a fertilized egg. For instance (as mentioned in chapter 3 and noted again here), all active organisms above a tenth or a hundredth of a millimeter in length have some system to propel fluids within their bodies. That's a strong signal pointing to the severe distance limitations of diffusion. And for instance—to take a more complex situation mentioned several times in this book—forests have always topped out, however favorable the circumstances, at something at or a little below a hundred meters in height. Some extrabiological limit or cost-benefit crossing point has been hit. Physical limits clearly constrain nature's diversity.

We can look, as well, at patterns in the diversity of the life that's present on the earth at the present time. A lot of work has focused on the significance of those patterns and on some of the puzzles they pose, work that asks in particular about what promotes and what inhibits the emergence of diversity in different places. I'll just note one pattern that's immediately relevant to the main themes here. Tropical rain forests represent the most diverse of all terrestrial habitats, with spectacular numbers of different species of trees living side by side, and with even more diverse insects and spiders in and among them. Deserts have far less diversity, as do the boreal forests of high latitudes. We deforest the tropics for fuelwood and to clear land for agriculture, but so diverse are its trees that we harvest only a few high-value woods from them. Aside from plantation forests, where trees grow as a crop (as where I live, in the southeastern United States), we get the bulk of our timber for building and pulp for papermaking from those less diverse northern forests. We take advantage of their lack of diversity, of their extensive natural stands of single species of trees.

What do I make of the high diversity in places where life faces no long-term shortages of water and extremes of temperature, and of the low diversity where life bumps up against such problems? Nothing particularly profound or original, but the message that such physical factors may be more constraining than the biological ones of competition, predation, and the like. The individual organism or individual species may be more affected by the biological factors, but larger-scale ecological and evolutionary processes generate much more diversity where the world is warm and wet than where it's cold and dry. In short, physical matters matter—a lot.

To which I immediately add a note of caution: almost all our environmental measurements record average conditions over time, whether over a millisecond or an epoch. All these physical variables vary—if a factor didn't vary, then we could measure it once and for all. For two reasons, that greatly complicates figuring out how physical factors affect (or afflict) living things. For one thing, extremes can be at least as influential as averages in determining who can live where and, more particularly, who wins where in the perpetual competition for a place in the sun. For another, what constitutes an extreme depends on the relevant time scale, and that varies as widely as the fluctuations themselves. For a leaf in the sun, temperature varies second to second, as in figure 8.5, so it's affected by brief shifts in air movement and cloud cover. For a tree that in the spring must raise its sap and will for decades defer reproduction in favor of growth, longer time scales for temperature extremes might be relevant. That the leaf is part of the tree adds yet another complication. The problem of time scale has been raised,[2] but little has as yet been done to resolve it. Quite likely we'll find no general answer, and we'll just have to live with still another group of variables, dealing with them through analyses that combine physiology and ecology in mixes that vary on almost case-by-case bases.

What Might Matter in Our Human World?

Has this been a book about pure science, or would applied science be more like it? One can easily envision applications of much that's here, whether for improving agriculture or for designing biomimetic devices. Whatever its practical implications, for me the account itself remains a piece of pure science. I don't mean by this to cast applied science as any lesser endeavor. But I have to admit an increasing antipathy toward conflating science with applied science. In its commonest form, that conflation serves to justify scientific work with the allure of alleviating our ills—corporeal, social, environmental, technological. True enough, science can play a helpful, even a critical, role in dealing with contemporary problems and affording us better technologies. But saying that all too often translates into tacit assumptions about motivation. The assumptions become more than merely tacit when we're pressed to declare a socially meritorious objective in order to obtain funding or to mollify administrators. I don't worry about applications when picking my projects—I'm more concerned about the generality of the hoped-for outcome and the intrinsic fascination of the issue under scrutiny. And my colleagues seem similarly motivated.

You can read a book such as this one with any of several viewpoints in mind. Direct biomimetic application may be the predominant fascination de jour. So as we look at leaves, we might consider mechanisms to realize and advantages to be gained—cooler roofs, cars, and roads, and less urban heating—from developing surfaces that reflect the near infrared. Or we may look at leaves for more general hints of opportunities we rarely exploit: drag reduction through flexible, reconfiguring structures, putting liquids under tension, or superhydrophobicity. Or we could treat leaves as biological exemplars, cases that help us unravel the mechanisms by which organisms deal with ice formation or local mechanical damage. There's a spectrum from immediate application to scientific generalization, a spectrum along which we shift according to personal preference or simple opportunism.

I think there's a stronger argument than simple intellectual honesty for keeping immediate applications at arms' length, perhaps even for worrying less in an immediate sense about discerning general scientific rules. It might be introduced with a quote from a socially concerned physicist and Nobel laureate, Dennis Gabor (1900–1979):

> The most important and urgent problems of the technology of today are no longer the satisfactions of the primary needs or of archetypal wishes, but the reparation of the evils and damages wrought by the technology of yesterday.[3]

So here's a final venting—or rant. We need to maintain a healthy scientific establishment at least as much as we need to maintain well-equipped companies of firefighters, and certainly more than we need to maintain peacetime armies. (The latter can too easily be co-opted by unscrupulous leaders for nefarious ventures.) Many of the problems we humans will surely face in the future, those evils and damages to which Gabor alluded, will yield only to science-based solutions. We can't call a scientific establishment into being on short, or even fairly long, notice. We either have it or we don't. That's the case whether "we" means a particular nation or all humanity, so globally integrated is contemporary science. We can train a bus driver in fairly short order if we have buses, an airplane pilot in a little more time if we have airplanes, even a doctor in a couple of years if we have hospitals.

But we don't train scientists so much as we bring them into participation in a preexisting scientific establishment, an establishment that lacks the immediate utility of buses, airplanes, and hospitals. Science represents more than a growing body of data or a technology; it's a way of looking at the world. Its outlook didn't come easily to humans—perhaps our mental wiring prefers revelation—at least when judged by its slow emergence in only one culture during one historical period. So, while the aspiring scientist may serve no apprenticeship directed toward normal occupational skills, an analogous process of acculturation represents an almost absolute necessity.

Even if what we do is of no social benefit, and even if we could state with confidence that our present activities will never confer that, science should be supported as a resource of people and facilities available for mobilization. Scientific competence can be maintained only inasmuch as it's exercised, being something much closer to a stable of horses than to a fleet of trucks. In the meantime, all of us can enjoy (or ought to enjoy) science as a source of enlightenment and aesthetic pleasure no less than what we find in music, art, and literature. This last sentiment motivates this book—I do hope you feel that it has been realized.

I've tried to stay with the symbols most commonly used. Since we're sticking our toes into a wide range of areas with varying conventions, some symbols represent different variables in different chapters. Similarly, some variables are represented by different symbols in different chapters. These inconsistencies ensure that if you pursue any item here beyond the present book, you'll use the conventional symbol.

See also http://physics.nist.gov/cuu/Units/units.html.

Variables

SYMBOL/REPRESENTS		SI UNITS
A	area	square meters
a	acceleration	meters/second^2
C	concentration	kilograms/meter^3
°C	temperature, Celsius	degrees above 273 K
C_d	drag coefficient	[none]
C_{stress}	stress concentration factor	[none]
c	speed of light	meters/second
D	diffusion coefficient	meters^2/second
D	drag	newtons
d	depth (of a beam)	meters
E	stiffness or Young's modulus	newtons/meter^2
e	2.71828	[none]
F	force	newtons
F_E	critical Euler buckling force	newtons
F_W	gravitational loading force	newtons
f	frequency	1/seconds (hertz)
Gr	Grashof number	[none]
g	gravitational acceleration	meters/second^2
h	height (distance upward)	meters
I	light intensity	candela
K	temperature, kelvin	Kelvin (absolute) degrees
l	length	meters
m	mass	kilograms
n	number of anything	[none]
p	pressure, cohesion	newtons/meter^2 (pascals)
$Pé$	Péclet number	[none]

Q	volume flow	meters^3/second
q	energy flow rate (power)	joules/second or watts
R	universal gas constant	joules/mole
Re	Reynolds number	[none]
r	radius	meters
S	area	square meters
T	temperature	Kelvin degrees
T	tension	newtons/meter
t	time	seconds
V	volume	meters^3
v	velocity	meters/second
w	width (of a beam)	meters
x	distance with flow or diffusion	meters
y	distance across a flow or load	meters
z	distance across a flow	meters
α	absorption coefficient	[none]
β	thermal expansion coefficient	1/degrees
γ	surface tension	newtons/meter
τ	shear stress	force/area
Δ	change in value of anything	[none]
δ	boundary layer thickness	meters
ε	radiant emissivity	[none]
ε	energy dissipation rate	joules/kilogram-second
Θ	contact angle	degrees
λ	wavelength	meters
μ	coefficient of viscosity	pascal-second
π	3.1416	[none]
ρ	density	kilograms/cubic meter
σ	Stefan-Boltzmann constant	watts/meter^{2}kelvin^4
σ	tensile stress	newtons/meter^2

Special Abbreviations

CAM	crassulacean acid metabolism
f	focal length
SI	Système Internationale (for units)
WUE	water use efficiency
$\propto$	"is proportional to"

Conversions to SI

To get from SI for all except temperature, divide instead of multiply by the factor given.

degrees Celsius = (degrees Fahrenheit – 32) × $\frac{5}{9}$

kelvins = degrees Celsius + 273.15

pascals (newtons/square meter) = atmospheres × 101,000

pascals (newtons/square meter) = millimeters of mercury × 133.3

meters/second = kilometers/hour × 0.278

meters/second = miles/hour × 0.447

kilograms = pounds mass × 0.454

newtons = pounds force × 4.45

meters = feet × 0.305

square meters = square feet × 0.0929

cubic meters = cubic feet × 0.0283

joules = kilocalories × 4187 = calories × 4.187

watts (joules/second) = kilocalories/hour × 1.163

Chapter One

1. Vogel (1996).
2. Vogel (1992).
3. Niklas (1997).
4. Thermodynamics, the second law in particular, is as daunting as it is important. Good introductions are Percy Bridgman's classic, *The Nature of Thermodynamics*, and P. W. Atkins's *Second Law*.
5. While I mentioned—and drew a picture of—Bénard cells in my book, *Life in Moving Fluids*, their use as an example of order generated by heat flow comes from P. C. W. Davies (1989). Ball (1999) provides a swell discussion of the phenomenon, with lots of pictures, in a delightfully engaging book.

Chapter Two

1. Anne Benninghoff, University of Michigan, drew my attention to properly quantitative work on rhododendrons. The main source here is Nilsen (1990).
2. Nobel (2005), not elementary but especially good on the physical side.
3. Chazdon (1988), which gives a good general picture of sunflecks.
4. Horn (1971), which tickles my adaptationist biases.
5. Burns (1975), a delight, especially "Biomodels," p. 79.
6. Niklas (1997), Ennos (2001). The last is a particularly engaging account for the general reader. While the tactical factors that determine the limits as tree height evolves in one lineage or another are matters of dispute, competition for access to light as the driver seems uncontroversial. See, for instance, King (1990) and Falster and Westoby (2003).

Chapter Three

1. Vogel (1994a). It was stimulated by an earlier piece in the journal, ostensibly about a demonstration of diffusion that asked students to note the swirling of dye dropped into a liquid. Swirling!!!
2. Berg (1993). If a book on diffusion can be said to be both sophisticated and engaging, this is it.
3. Most accounts of the so-called Manhattan Project look mainly at the Los Alamos laboratory and the physics involved. By contrast, Groueff (1967) gives a view of the remarkable engineering involved, including isotope separation systems.
4. If you want a more complete treatment of this odd business, Monteith and Unsworth (2008) does a good job. Berg (1993) touches on it as well. The original reference, I think, is Brown and Escomb (1900).

5. Miles Silman and Bill Smith of Wake Forest University drew my attention to this peculiar difference between plants and people.

Chapter Four

1. A book I wrote some years ago (Vogel 1994b) can tell you more than you probably want to know about fluid flow in biology.
2. Vogel (2009a). The first chapter is all about the Péclet number.

Chapter Five

1. Traditionally obtained from tabulated values, these days conveniently calculated on demand by the e-folks at www.ringbell.co.uk/info/humid.htm.
2. Vogel (2001). The datum was first determined at the dawn of the industrial revolution, not surprisingly. It wasn't expressed in watts—James Watt was born about when it was first measured.
3. According to Service (2009).
4. Sinclair et al. (1984) provide a good introduction to water-use efficiency. A whole lot more about the subject can be found in the book edited by Bacon (2004). Nobel (2005) puts the whole business in context.
5. Sinclair (2009).
6. Good general sources on stomata are the books by Martin et al. (1983) and Willmer and Fricker (1996).
7. Any textbook of animal physiology will provide the details; I particularly like Schmidt-Nielsen (1997). The same case is made in the ninth chapter of a book I did recently (2009a).

Chapter Six

1. In *Vegetable Staticks* (1727), Hales describes his various experimental manipulations; it's still fairly easy to find and still good reading.
2. Chapotin et al. (2006).
3. Dixon and Joly (1895).
4. Briggs (1950). Lawrence Briggs took on the project upon retirement as head of the US National Bureau of Standards.
5. I got the story from Sprackling's (1985) lovely little book, *Liquids and Solids*, one in a series (Student Physics) for first-year physics undergraduates in the UK. Brennen (1995) mentions this and quite a few other, later, measurements—with proper references.
6. See, for instance, Wilson et al. (1975) and Brennen (1995).
7. Scholander et al. (1965) is the primary source.
8. Holbrook et al. (1995); Pockman et al. (1995).
9. Sachs (1882), as translated by H. Marshall Ward in 1887.

10. Zimmermann and Brown (1971) have collected the old data.
11. The general argument for safety (against embolisms) versus fluid conductivity has been put in far more sophisticated terms by Comstock and Sperry (2000).

Chapter Seven

1. Robinson (2005) provides a most engaging (quite beyond a thirty-two-word title) account of the life and work of Thomas Young.
2. Tyree and Zimmermann (2002) give more detail as well as a clear general account.
3. See, for instance (and a good read), Cronon (1983).
4. Jansen et al. (2004) give a good account of these so-called vestured pits.
5. A summary of the state of the art is given by Clearwater and Goldstein (2005). I can't resist mentioning the title of a paper by two of the best workers, Holbrook and Zwieniecki (1999): "Embolism Repair and Xylem Tension: Do We Need a Miracle?"
6. See, for a good account of the issues and literature, Tyree and Zimmermann (2002).
7. For present purposes, Nobel (2005) is particularly informative. A more general source is P. W. Ford's contribution (pp. 686–92) in Chesworth (2008).
8. Fitter and Hay (2002) summarize the situation. The classic work on the roots of mangroves is that of the versatile Per Scholander, the bomber mentioned in the previous chapter (Scholander et al. 1955).

Chapter Eight

1. Self-reference may be a sure sign of scientific narcissism, but I do want to mention two recent papers. In one (Vogel 2005) I give a brief introduction to the modes of heat transfer that matter to organisms. In the other (Vogel 2009b) I summarize what we know about how leaves respond to the danger of excessive thermal load.
2. While the calculations are mine, most of the data from which I start come from Nobel (2005).
3. Datum from Vogel (1984a).
4. I get mine from Edmund Scientific, Tonowanda, NY (www.scientificsonline.com).
5. Ehleringer et al. (1976).

Chapter Nine

1. I've taken the names and a lot of the story I tell from an excellent review of both the diversity of superhydrophobic plant surfaces and the way superhydrophobicity works by Koch et al. (2008).

2. Cassie and Baxter (1944) were concerned with water repellency of fabrics, but mention with admiration the facility of ducks' outer feathers to shed water and the contribution of the feathers' roughness.

3. Barthlott and Neinhuis (1997).

4. Colmer and Pedersen (2008).

5. Thorpe and Crisp (1947). Earlier work of others certainly pointed in that direction, but they worked out the mechanism.

6. Lücking and Bernecker-Lücking (2005).

7. Burd (2007).

8. Gosnell (2005) relates more than you'd have thought you'd enjoy knowing about ice.

9. The fully nuanced story can be found in Vogel (1994b).

10. Aylor and Parlange (1975) seem to have been the first to look into the matter. A more elaborate analysis (although specifically concerned with pollen shedding) is given by Urzay et al. (2009).

11. I'm mainly relying on two sources, a paper by Press (1999) and a textbook by Gates (1980).

12. Egri et al. (2010).

13. Wright and Westoby (2002) put the matter in nice perspective.

Chapter Ten

1. Wharton (2002), who did the work on nematodes, provides a fine general account of how creatures cope with extremes.

2. Sakai and Larcher (1987) is my main source for how plants respond to freezing temperatures. Smallwood and Bowles (2002) and Griffith and Yaish (2004) provide updates, particularly for antifreeze proteins.

3. Garrett Hardin (1956) wrote a wonderfully iconoclastic essay, "Meaningless of the Word Protoplasm." The word did persist for a while in introductory textbooks, always loath to delete any term that might provide easily graded grist for an exam.

4. The case is reviewed by Guy (2003).

5. Wolfe and Bryant (1999); Ball et al. (2002).

6. Gosnell (2005), referred to in the previous chapter, gives the story.

7. See, for instance, Rall and Fahy (1985) and Liu et al. (2008).

8. Sformo et al. (2010).

9. Scholander et al. (1953).

Chapter Eleven

1. These and other data in the chapter come from the tabulated values I collected for a textbook (Vogel 2003).

2. The classic that lives above my desk is Oberg et al. (1987), the twenty-second edition of a compendium that first appeared in 1914. That's shelf life!

3. Copeland (1866, pp. 364–69) sings their praises for ordinary greenhouses, noting the way the increased availability of iron allows the construction of frames that block much less light than would the equivalent in timber, and making proper reference to Sir Joseph Paxton.

4. Greenhill (1881). A. G. Greenhill contributed to quite a range of problems in mathematical physics. He's perhaps best known for calculating how rapidly a bullet should spin as it passes down the barrel of a rifle. (We rarely recall that the word *rifle* refers to the grooving or rifling within a barrel that imparts spin, the latter persuading the bullet not to tumble in flight.)

5. Thompson (1942). Thompson's book, originally written in 1917, definitively re-written in 1942, makes glorious reading, especially in the Bonner abridgment of 1961. Its hyper-Victorian prose and wealth of ideas has kept it conspicuous, but it's too easy to miss its oddly anachronistic viewpoint—I've met nonbiologists who think it reflects our current thinking. Thompson was no Darwinian or even Aristotelian. Instead, he was searching for a kind of Pythagorean mathematical perfection in nature.

Chapter Twelve

1. Frazer (1962). The data were presented again by Mayhead (1973) in a less obscure publication.

2. Vogel (1984b). Yes, I had let things go until the last minute, submitting an ad-equately vague abstract earlier. It's not a practice I recommend. At least the fin-ished manuscript wasn't due right at presentation time.

3. Vogel (1989). During the normal review process, the manuscript received some curious and amusing comments. The reviewers, it appeared, were as surprised as I had been that something so easy, interesting, and functionally relevant had not been done by someone long ago. Each quite obviously (and quite properly) plumbed the literature and came up dry, so I became more confident that I had done my homework.

4. Niklas (1999)—a particularly nice review of the mechanics of leaves.

5. Ibid.

6. See, for instance, Givnish (1978).

7. May (1998).

8. Roland Ennos (2001) has written a beautiful (both text and illustrations) popular book that provides a good picture of (among other things) the mechanical prob-lems of being a tree.

9. Mickovski and Ennos (2003) slowly winched over pines (Macedonian pines, *Pinus peuce*) while making careful observations of what happened.

10. The best papers on the general mechanics of buttresses, I think, are those of Ennos (1993) and Crook et al. (1997).

11. Such as that of Baker (1995).

Chapter Thirteen

1. Crane (1950).
2. Raup (1962; 1966 on mollusk shells).
3. Kobayashi et al. (1998). The work is also described in a popular book on biomimetics by Peter Forbes (2005). Mahadevan and Rica (2005) give the matter more general and theoretical treatment.
4. I vaguely recall hearing long ago that the Volkswagen Karmann-Ghia, a now-classic sporty coupe version of the Beetle with much-admired lines, required hand joining by welding and smoothing of some of the body panels. An Internet search suggests the same. Pricey? The car cost half again as much as the Beetle, its mechanical near twin.
5. My colleague Fred Nijhout has conducted a long and elegant series of investigations on how insects trigger molting and, eventually, pupation; they really do keep track of how big they are (Nijhout 1975 and other papers).
6. Katifori et al. (2010). At this writing some of their videos are accessible; see www.sciencefriday.com/videos/watch/10277.
7. This is the trichotomization suggested by Fitter and Hay (2002).
8. Massey et al. (2007) make this point.
9. My sources here are Wainwright et al. (1976) and Vincent (1990a).
10. A splendid introduction to the subject of fracture mechanics is Gordon (1976). An equally splendid book, a real page-turner, is his more general account (Gordon 1978).
11. Koehl et al. (2008) summarizes the situation for macroalgae.
12. A good review of tearing in plants, with special attention to grass leaves, where he first drew attention to their resistance and to its implications, is that of Vincent (1990b). A quick account of what grasses do when nicked can be found in Vincent (1990a).
13. My source on cellulose digestion and often-consulted physiology text is Schmidt-Nielsen (1997)—and not simply because he was a good friend and sometime-mentor.
14. Nowak (1991). The book, *Walker's Mammals of the World*, is a treasure of authoritative information.

Chapter Fourteen

1. The basic reference on scaling in plants is Niklas (1994). My favorite on scaling in general is McMahon and Bonner (1983).
2. In particular by Denny and Gaines (2000).
3. Gabor (1970), p. 9.

Atkins, P. W. 1984. *The Second Law*. New York: Scientific American Books. (12, 281)

Aylor, D. E., and J.-Y. Parlange. 1975. "Ventilation Required to Entrain Small Particles from Leaves." *Plant Physiol.* 56: 97–99. (175, 284)

Bacon, M. A., ed. 2004. *Water Use Efficiency in Plant Biology*. Oxford: Oxford University Press. (75, 282)

Baker, C. J. 1995. "The Development of a Theoretical Model for the Windthrow of Plants." *J. Theor. Biol.* 175: 355–72. (237, 285)

Ball, M. C., J. Wolfe, M. Canny, M. Hofmann, A. B. Nicotra, and D. Hughes. 2002. "Space and Time Dependence of Temperature and Freezing in Evergreen Leaves." *Funct. Plant Biol.* 29: 1259–72. (190, 284)

Ball, P. 1999. *The Self-Made Tapestry: Pattern Formation in Nature*. New York: Oxford University Press. (15, 281)

Barthlott, W., and C. Neinhuis. 1997. "Purity of the Sacred Lotus, or Escape from Contamination in Biological Surfaces." *Planta* 202: 108. (166, 284)

Berg, H. C. 1993. *Random Walks in Biology*. Exp. ed. Princeton, NJ: Princeton University Press. (49, 53, 281)

Brennen, C. E. 1995. *Cavitation and Bubble Dynamics*. New York: Oxford University Press. (101, 102, 282)

Bridgman, P. W. 1941. *The Nature of Thermodynamics*. Cambridge, MA: Harvard University Press. (12, 281)

Briggs, L. J. 1950. "Limiting Negative Pressure of Water." *J. Appl. Phys.* 21: 721–22. (99, 282)

Brown and Escomb. 1900. "Static Diffusion of Gases and Liquids in Relation to the Assimilation of Carbon and Its Translocation in Plants." *Phil. Trans. Roy. Soc. Lond.* B193: 223–91. (53, 281)

Burd, M. 2007. "Adaptive Function of Drip Tips: A Test of the Epiphyll Hypothesis in *Psychotria marginata* and *Faramea occidentalis* (Rubiaceae)." *J. Trop. Ecol.* 23: 449–55. (172, 284)

Burns, J. M. 1975. *Biograffiti: A Natural Selection*. New York: Quadrangle/New York Times Book Company. (37, 281)

Cassie, A. B. D., and S. Baxter. 1944. "Wettability of Porous Surfaces." *Trans. Faraday Soc.* 40: 546–551. (166, 284)

Chapotin, S. M., J. H. Razanameharizaka, and N. M. Holbrook. 2006. "Baobab Trees (Adansonia) in Madagascar Use Stored Water to Flush New Leaves but Not to Support Stomatal Opening before the Rainy Season." *New Phytol.* 169: 549–59. (95, 282)

Chazdon, R. L. 1988. "Sunflecks and Their Importance to Forest Understory Plants." *Adv. Ecol. Res.* 18: 1–63. (35, 281)

Chesworth, W., ed. 2008. *Encyclopedia of Soil Science.* Dordrecht, Netherlands: Springer. (134, 283)

Clearwater, M. J., and G. Goldstein. 2005. "Embolism Repair and Long Distance Water Transport." In *Vascular Transport in Plants,* ed. N. M. Holbrook and M. A. Zwieniecki, pp. 375–99. Burlington, MA: Elsevier Academic Press. (133, 283)

Colmer, T. D., and O. Pedersen. 2008. "Underwater Photosynthesis and Respiration in Leaves of Submerged Wetland Plants: Gas Films Improve CO_2 and O_2 Exchange." *New Phytol.* 177: 918–26. (169, 284)

Comstock, J. P., and J. S. Sperry. 2000. "Theoretical Considerations of Optimal Conduit Length for Water Transport in Vascular Plants." *New Phytol.* 148: 195–218. (114, 283)

Copeland, R. M. 1866. *Country Life: A Handbook of Agriculture, Horticulture, and Landscape Gardening.* Boston, MA: Dinsmoor. (208, 285)

Crane, H. R. 1950. "Principles and Problems of Biological Growth." *Sci. Monthly* 70: 376–89. (239, 286)

Cronon, W. 1983. *Changes in the Land.* New York: Hill and Wang. (132, 282)

Crook, M. J., A. R. Ennos, and J. R. Banks. 1997. "The Function of Buttress Roots: A Comparative Study of the Anchorage Systems of Buttressed (*Aglaia* and *Hephelium ramboutan* Species) and Non-Buttressed (*Mallotus wrayi*) Tropical Trees." *J. Exp. Bot.* 48: 1703–16. (234, 285)

Davies, P. C. W. 1989. "The Physics of Complex Organisation." In *Theoretical Biology,* ed. B. Goodwin and P. Saunders, pp. 101–11. Edinburgh, UK: Edinburgh University Press. (15, 281)

Denny, M. W., and S. Gaines. 2000. *Chance in Biology: Using Probability to Explore Nature.* Princeton, NJ: Princeton University Press. (273, 286)

Dixon, H. H., and J. Joly. 1895. "On the Ascent of Sap." *Phil. Trans. Roy. Soc. Lond. B* 186: 563–76. (95, 282)

Egri, A., A. Horváth, G. Kriska, and G. Horváth. 2010. "Optics of Sunlit Water Drops on Leaves: Conditions under Which Sunburn Is Possible." *New Phytologist* 185: 979–87. (178, 284)

Ehleringer, J., O. Björkman, and H. A. Mooney. 1976. "Leaf Pubescence: Effects on Absorptance and Photosynthesis in a Desert Shrub." *Science* 192: 376–77. (157, 283)

Ennos, A. R. 1993. "The Function of Buttresses." *Tree* 8: 350–51. (234, 285)

———. 2001. *Trees.* London: Natural History Museum; Washington, DC: Smithsonian Institution Press. (38, 229, 281, 285)

Falster, D. S., and M. Westoby. 2003. "Plant Height and Evolutionary Games." *Trends Ecol. Evol.* 18: 337–43. (38, 281)

Fitter, A. H., and R. K. M. Hay. 2002. *Environmental Physiology of Plants.* 3rd ed. London: Academic Press. (139, 247, 283, 286)

Forbes, P. 2005. *The Gecko's Foot: Bioinspiration; Engineering New Materials from Nature.* New York: W. W. Norton. (244, 286)

Frazer, A. I. 1962. "Wind Tunnel Studies of the Forces Acting on the Crowns of Small Trees." *Rep. Forest Res.* (*UK*) 1962: 178–83. (218, 285)

Gabor, D. 1970. *Innovation: Scientific, Technological and Social*. Oxford: Oxford University Press. (274, 286)

Gates, D. M. 1980. *Biophysical Ecology*. Mineola, NY: Dover Publications. (177, 284)

Givnish, T. J. 1978. "On the Adaptive Significance of Compound Leaves, with Particular Reference to Tropical Trees." In *Tropical Trees as Living Systems*, ed. P. B. Tomlinson and M. H. Zimmermann, pp. 351–80. Cambridge, UK: Cambridge University Press. (225, 285)

Gordon, J. E. 1976. *The New Science of Strong Materials*. Harmondsworth, UK: Penguin Books. (253, 286)

———. 1978. *Structures; or, Why Things Don't Fall Down*. New York: Plenum Press. (253, 286)

Gosnell, M. 2005. *Ice: The Nature, the History, and the Uses of an Astonishing Substance*. New York: Alfred A. Knopf. (173, 190, 284)

Greenhill, A. G. 1881. "Determination of the Greatest Height Consistent with Stability That a Vertical Pole or Mast Can Be Made, and of the Greatest Height to Which a Tree of Given Proportions Can Grow." *Cambridge Phil. Soc.* 4: 65–73. (212, 285)

Griffith, M., and M. W. F. Yaish. 2004. "Antifreeze Proteins in Overwintering Plants: A Tale of Two Activities." *Trends Plant Sci.* 9: 399–405. (188, 284)

Groueff, S. 1967. *Manhattan Project: The Untold Story of the Making of the Atomic Bomb*. Boston: Little, Brown. (49, 281)

Guy, C. L. 2003. "Freezing Tolerance of Plants: Current Understanding and Selected Emerging Concepts." *Can J. Bot.* 81: 1216–23. (189, 284)

Hales, S. 1727. *Vegetable Staticks*. London: W. & J. Innys and T. Woodward. Reprint, London: Scientific Book Guild, 1961. (94, 282)

Hardin, G. 1956. "Meaninglessness of the Word Protoplasm." *Sci. Monthly* 82: 112–20. (188, 284)

Holbrook, N. M., M. J. Burns, and C. B. Field. 1995. "Negative Xylem Pressures in Plants: A Test of the Balancing Pressure Technique." *Science* 270: 1193–94. (106, 282)

Holbrook, N. M., and M. A. Zwieniecki. 1999. "Embolism Repair and Xylem Tension: Do We Need a Miracle?" *Plant Physiol.* 120: 7–10. (106, 133, 283)

Horn, H. S. 1971. *The Adaptive Geometry of Trees*. Princeton, NJ: Princeton University Press. (36, 281)

Jansen, S., P. Baas, P. Gasson, F. Lens, and E. Smets. 2004. "Variation in Xylem Structure from Tropics to Tundra: Evidence from Vestured Pits." *Proc. Natl. Acad. Sci. USA* 101: 8833–37. (132, 283)

Johnson, S. 2008. *The Invention of Air*. New York: Riverhead Books. (18)

Katifori, E., G. J. Szöllósi, and M. O. Magnasco. 2010. "Damage and Fluctuations Induce Loops in Optimal Transport Networks." *Phys. Rev. Letts.* 104: 1–4, doi:10.1103/104.048704. (246, 286)

King, D. A. 1990. "The Adaptive Significance of Tree Height." *Amer. Nat.* 135: 809–28. (38, 281)

Kobayashi, H., B. Kresling, and J. F. V. Vincent. 1998. "The Geometry of Unfolding Tree Leaves." *Proc. R. Soc. Lond.* B265: 147–54. (244, 286)

Koch, K., B. Bhushan, and W. Barthelott. 2008. "Diversity of Structure, Morphology and Wetting of Plant Surfaces." *Soft Matter* 4: 1943–63. (165, 283)

Koehl, M. A. R., W. K. Silk, H. Liang, and L. Mahadevan. 2008. "How Kelp Produce Blade Shapes Suitable for Different Flow Regimes: A New Wrinkle." *Integr. Comp. Biol.* 48: 834–51. (256, 286)

Liu, B., J. J. McGrath, and B. Wang. 2008. "Determination of the Ice Quantity by Quantitative Microscopic Imaging of Vitrifying Solutions." *Biopreservation and Biobanking* 6: 261–68. (194, 284)

Lücking, R., and A. Bernecker-Lücking. 2005. "Drip-Tips Do Not Impair the Development of Epiphyllous Rain-Forest Lichen Communities." *J. Trop. Ecol.* 21: 171–77. (170, 284)

Mahadevan, L., and S. Rica. 2005. "Self-Organized Origami." *Science* 307: 1740. (244, 286)

Martin, E. S., M. E. Donkin, and R. A. Stevens. 1983. *Stomata*. London: Edward Arnold. (81, 282)

Massey, F. P., A. R. Ennos, and S. E. Hartley. 2007. "Grasses and the Resource Availability Hypothesis: The Importance of Silica-Based Defences." *J. Ecol.* 95: 414–24. (248, 286)

May, M. 1998. "Leaves Are Such a Drag." *Muse* 2, no. 1: 10–14. (226, 285)

Mayhead, G. I. 1973. "Some Drag Coefficients for British Forest Trees Derived from Wind Tunnel Studies." *Agric. Meteorol.* 12: 123–30. (218, 285)

McMahon, T. A., and J. T. Bonner. 1983. *On Size and Life*. New York: Scientific American Books. (265, 286)

Mickovski, S. B., and A. R. Ennos. 2003. "Anchorage and Asymmetry in the Root System of *Pinus peuce*." *Silva Fennica* 37: 161–73. (232, 285)

Monteith, J. L., and M. U. Unsworth. 2008. *Principles of Environmental Physics*. 3rd ed. Burlington, MA: Academic Press. (33, 281)

Nijhout, H. F. 1975. "Threshold Size for Metamorphosis in Tobacco Hornworm, *Manduca sexta* (L.)." *Biol. Bull.* 149: 214–25. (246, 286)

Niklas, K. J. 1994. *Plant Allometry*. Chicago: University of Chicago Press. (286, 265)

———. 1997. *The Evolutionary Biology of Plants*. Chicago: University of Chicago Press. (10, 38, 281)

———. 1999. "A Mechanical Perspective on Foliage Leaf Form and Function." *New Phytol.* 143: 19–31. (224, 285)

Nilsen, E. T. 1990. "Why Do Rhododendron Leaves Curl?" *Arnoldia* 50: 30–35. (28, 281)

Nobel, P. S. 2005. *Physicochemical and Environmental Plant Physiology*. 3rd ed. Burlington, MA: Elsevier. (24, 33, 75, 134, 141, 281, 282, 283)

Nowak, R. M. 1991. *Walker's Mammals of the World*. 5th ed. Baltimore: Johns Hopkins University Press. (261, 286)

Oberg, E., F. D. Jones, and H. L. Horton. 1984. *Machinery's Handbook*. 22nd ed. New York: Industrial Press. (205, 284)

Pockman, W. T., J. S. Sperry, and J. W. O'Leary. 1995. "Sustained and Significant Negative Water Pressure in Xylem." *Nature* 378: 715–16. (106, 282)

Press, M. C. 1999. "Functional Significance of Leaf Structure: A Search for Generalizations." *New Phytologist* 143: 213–19. (177, 284)

Priestley, J., and W. Hey. 1772. "Observations on Different Kinds of Air." *Phil. Trans. Roy. Soc. Lond.* 62: 147–264. (18)

Rall, W. F., and G. M. Fahy. 1985. "Ice-Free Cryopreservation of Mouse Embryos at –196° C by Vitrification." *Nature* 313: 573–75. (194, 284)

Raup, D. M. 1962. "Computer as Aid in Describing Form in Gastropod Shells." *Science* 138: 150–52. (239, 286)

———. 1966. "Geometric Analysis of Shell Coiling: General Problems." *J. Paleontol.* 40: 1178–90. (239, 286)

Robinson, A. 2005. *The Last Man Who Knew Everything: Thomas Young, the Anonymous Polymath Who Proved Newton Wrong, Explained How We See, Cured the Sick, and Deciphered the Rosetta Stone, among Other Feats of Genius.* New York: Pi Press. (118, 283)

Sachs, J. von. 1882. *Lectures on the Physiology of Plants*. Translated by H. M. Ward, Oxford: Clarendon Press, 1887. (107, 282)

Sakai, A., and W. Larcher. 1987. *Frost Survival of Plants*. Berlin: Springer-Verlag. (188, 284)

Schmidt-Nielsen, K. 1997. *Animal Physiology: Adaptation and Environment*. 5th ed. Cambridge, UK: Cambridge University Press. (90, 260, 282, 286)

Schofield, R. E. 1997. *The Enlightenment of Joseph Priestley: A Study of His Life and Work from 1733 to 1773.* University Park: Pennsylvania State University Press. (18)

———. 2004. *The Enlightened Joseph Priestley: A Study of His Life and Work from 1773 to 1804.* University Park: Pennsylvania State University Press. (18)

Scholander, P. F., W. Flagg, R. J. Hock, and L. Irving. 1953. "Studies on the Physiology of Frozen Plants in the Arctic." *J. Cell. Comp. Physiol.* 42, Suppl. 1: 156. (196, 284)

Scholander, P. F., H. T. Hammel, E. D. Bradstreet, and E. A. Hemmingson. 1965. "Sap Pressure in Vascular Plants." *Science* 148: 339–46. (104, 139, 282)

Service, R. R. 2009. "Another Biofuels Drawback: The Demand for Irrigation." *Science* 326: 216–17. (74, 282)

Sformo, T., K. Walters, K. Jeannet, B. Wowk, G. M. Fahy, B. M. Barnes, and J. G. Duman. 2010. "Deep Supercooling, Vitrification and Limited Survival to -100° C in the Alaskan Beetle *Cucujus clavipes puniceus* (Coleoptera: Cucujidae) Larvae." *J. Exp. Biol.* 213: 502–9. (194, 284)

Sinclair, T. R. 2009. "Taking Measure of Biofuels Limits." *Amer. Sci.* 97: 400–407. (77, 282)

Sinclair, T. R., C. B. Tanner, and J. M. Bennett. 1984. "Water-Use Efficiency in Crop Production." *BioScience* 34: 36–40. (75, 282)

Smallwood, M., and D. J. Bowles. 2002. "Plants in a Cold Climate." *Phil. Trans. R. Soc. Lond.* B357: 831–47. (188, 284)

Sprackling, M. T. 1985. *Liquids and Solids.* London: Routledge and Kegan Paul. (101, 282)

Thompson, D'Arcy W. 1942. *On Growth and Form.* 2nd ed. Cambridge: Cambridge University Press. (212, 285)

Thorpe, W. H., and D. J. Crisp. 1947. "Studies on Plastron Respiration. I. The Biology of *Aphelocheirus* [Hemiptera, Aphelocheiridae (Nauchoridae)] and the Mechanism of Plastron Retention." *J. Exp. Biol.* 24: 227–69. (169, 284)

Tyree, M. T., and M. H. Zimmermann. 2002. *Xylem Structure and the Ascent of Sap.* Berlin: Springer. (126, 133, 283)

Urzay, J., S. G. Llewellyn Smith, E. Thompson, and B. J. Glover. 2009. "Wind Gusts and Plant Aeroelasticity Effects on the Aerodynamics of Pollen Shedding: A Hypothetical Turbulence-Initiated Wind-Pollination Mechanism." *J. Theor. Biol.* 259: 785–92. (175, 284)

Vincent, J. F. V. 1990a. *Structural Biomaterials.* Rev. ed. Princeton, NJ: Princeton University Press. (250, 259, 286)

———. 1990b. "Fracture Properties of Plants." *Adv. in Bot. Res.* 17: 235–87. (259, 286)

Vogel, S. 1984a. "The Thermal Conductivity of Leaves." *Can. J. Bot.* 62: 741–44. (141, 283)

———. 1984b. "Drag and Flexibility in Sessile Organisms." *Amer. Zool.* 24: 37–44. (219, 285)

———. 1989. "Drag and Reconfiguration of Broad Leaves in High Winds." *J. Exp. Bot.* 40: 941–48. (222, 285)

———. 1992. *Vital Circuits.* New York: Oxford University Press. (10, 281)

———. 1994a. "Dealing Honestly with Diffusion." *Amer. Biol. Teacher* 56: 405–7. (43, 281)

———. 1994b. *Life in Moving Fluids.* Princeton, NJ: Princeton University Press. (15, 58, 175, 281, 282, 284)

———. 1996. "Diversity and Convergence in the Study of Organismal Function." *Israel J. Zool.* 42: 297–305. (10, 281)

———. 2001. *Prime Mover.* New York: W. W. Norton. (72, 282)

———. 2003. *Comparative Biomechanics: Life's Physical World.* Princeton, NJ: Princeton University Press. (202, 284)

———. 2005. "Living in a Physical World. IV. Moving Heat Around." *J. Biosci.* 30: 449–60. (141, 283)

———. 2009a. *Glimpses of Creatures in Their Physical Worlds.* Princeton, NJ: Princeton University Press. (90, 282)

———. 2009b. "Leaves in the Lowest and Highest Winds: Temperature, Force and Shape." *New Phytol.* 183: 13–26. (69, 141, 283)

Wainwright, S. A., W. D. Biggs, J. D. Currey, and J. M. Gosline. 1976. *Mechanical Design in Organisms.* London: Edward Arnold. (250, 286)

Wharton, D. A. 2002. *Life at the Limits.* Cambridge: Cambridge University Press. (188, 284)

Wigner, E. P. 1960. "The Unreasonable Effectiveness of Mathematics in the Natural Sciences." *Comm. Pure Appl. Math.* 13: 1–14. (5)

Willmer, C., and M. Fricker. 1996. *Stomata.* 2nd ed. London: Chapman & Hall. (81, 282)

Wilson, D. A., J. W. Hoyt, and J. W. McKune. 1975. "Measurement of Tensile Strength of Liquids by an Explosion Technique." *Nature* 253: 723–25. (102, 282)

Wolfe, J., and G. Bryant. 1999. "Freezing, Drying, and/or Vitrification of Membrane-Solute-Water Systems." *Cryobiology* 39: 103–29. (190, 284)

Wright, I. J., and M. Westoby. 2002. "Leaves at Low versus High Rainfall: Coordination of Structure, Lifespan and Physiology. *New Phytologist* 155: 403–16. (178, 284)

Zimmermann, M. H., and C. L. Brown. 1971. *Trees: Structure and Function.* New York: Springer-Verlag. (113, 283)

Note: Italicized page numbers indicate figures.

center of gravity, 231, 232

chlorophyll, 24

climate change, 195

cohesion vs adhesion, 122, 128

cohesion-tension theory. *See* pressure, negative

colligative properties, 84, 184, 192. *See also specific properties*

columns, 210–13

compensation point, 33

composite materials, 198

compression: compressive crushing, 210; gases, 199; location in cantilever, 204; location in leaf, 204; resistance by water, 199; vs tension, 198; withstanding, 204

conduction of heat, 146

conduits. *See* xylem

contact angle, 164, 165

control systems, 86

convection, 15, 145–57, 272; baking pizza, 153; vs diffusion, 42; forced, 152–56; free, 147–52, 154; free vs forced, 146; limitation by hairiness, 177; mixed (*see* convection, forced); pizza, 152; in plant cells, 51, 67; rhododendron leaves, 28; and thermal conductivity, 149, 151. *See also* flow

convergence, 9–11; hairy leaves, 176; hearts, 10; and multifunctionality, 176; tree height, 10

cooling: convection, 145–57; evaporation, 144–45; leaves, 140–61; reradiation, 141–43; speed of, 188, 193; supercooling, 193

cracks: energy to propagate, 189; of ice, 173. *See also* fracture

Crane, Horace, 239

crassulacean acid metabolism, 278

curved surfaces, 116–24, 255

cyclosis, 68, 266

daffodil flowers, 241, 242

deformation. *See* leaves, reconfiguration

density, 278; air, 217; ice-melting gradient, 182; water, 144, 172, 195, 217; wood, 211, 229

diffusion, 41–55, 56; vs altitude, 53–54; in beverages, 130; carbon dioxide, 65, 72, 77, 79; coefficient, 42, 47, 52, 54, 72, 267, 277; CO_2 vs H_2O, 72–74; demonstrating, 48–49, 50; vs convection, 42; directionality, 44; vs distance, 266; Fick's law, 46–47, 53; vs flow, 60, 63–69; gases, 72, 90, 169; intracellular, 67; vs osmosis, 82; vs pubescence, 157; random walk, 44; "rate," 47–49; size and, 51, 49, 272; solubility vs, 90; stomatal stimulus, 88; thermodynamics of, 46; through membranes, 186; through stomata, 52, 79; water vapor, 42, 70, 71, 80; within leaves, 51, 79

dimensionless ratios, 64, 69, 146; Bond number, 168; Grashof number, 148, 154, 277; Péclet number, 64–69, 110, 277; Reynolds number, 110, 154, 217, 263, 278

dimensions, 48

distillation, 186

Dixon and Joly, 95, 97, 98, 103

drag, 215–17, 277; coefficient, 217, 220, 277; direction re flow, 216, 228; estimating, 227; flags, 218, 220, 224; leaves, 3, 217–27; minimization, 269; and Reynolds number, 217; speed dependence, 217, 218, 222; streamlining, 217; trees, 213, 215, 218, 269; weathervanes, 219, 224

drip tips, 10, 170–74, 171

droplets, 123–24; contact angle, 165; drip tips, 10, 170–74, 171; light focusing, 178; size, 171; on slope, 167, 168; superhydrophobicity, 167 efficiency, 13, 141

embolisms, 107, 127, 128, 133, 136

emissivity, 25, 278

energy, 11–16; conservation, 11, 147; folivory yield, 249, 259; in food, 103; lifting water, 71; making surface, 189, 253; moving fluids, 61; osmotic, 85, 139; for photosynthesis, 21; as pressure gradient, 109; rate of use, 278; release in freezing, 188; in sunlight, 20, 21, 24, 141, 144; for thermogenesis, 184; for xylem flow, 115

epiphytes, 171

Euler buckling, *210*, 211

evaporation, 71, 79, 84, 144–45; cooling by, 72, 144–45; vs humidity, 145; vs negative pressure, 106; and vapor pressure, 185

evolution, 6–11; adaptationism, 9; broad leaves, 62; convergent, 9–11, 78, 176, 265; creativity of, 8, 271; and diversity, 269, 272; fitness, 8; forests, 38–40; leaves, 62; light absorption, 22; limitations, 271, 272; and multifunctionality, 176; photosynthesis, 40

exaptations, 177

feedback, 86, *87*, 195.

fibers, 197

Fick's law, 46–47, 53

flow, 56–69; air vs water, 56–57; around leaf holes, 247; cyclosis, 68; vs diffusion, 60, 63–69; and drag, 216; Hagen-Poiseuille equation, 109; Kolmogorov scale, 60; laminar vs turbulent, 57, 58, 109, 263; near surfaces, 60–63, *108*; no-slip condition, 60–62, *108*; parabolic, *108*; Péclet number, 64–69; pressure loss from, 109; Reynolds number, 110, 154, 217, 263, 278; sap in xylem, 94, 111, 107–15, 138; scale of vortices, 59; viscosity, 57; wind

across leaves, 174. *See also* convection; fluids; velocity gradients; wind

flower color, 21

fluids; drag of moving, 215–17; gases and liquids as, 216. *See also* flow; gases; liquids

folivory. *See* herbivory

force, 277; components, 227; leverage, *216*. *See also* drag; gravity; pressure; support

forests: deciduous, 31; evolution, 38–40; floor light, 31, 35; hardwood vs softwood, 38; height, 38–40, 107, 138, 264; layers of leaves, 35–38; succession, 37; tropical, 32, 225, 232, 272; water use by, 145

fracture mechanics: brittle fracture, 253; crack stoppers, 258; toughness, 253

Franklin, Benjamin, 17

freezing: of cells, 201; energy release, 188; frost, *26*; of leaves, 191; nucleation, 188, 189; outgassing in, 131; point, 84, 184, 186, 195; prevention, 184–87; vegetables, *180*; volume increase, 190. *See also* ice

frequency vs wavelength, 19

frost. *See* freezing; ice

Gabor, Dennis, 274

gases: bubbles in xylem, 107; buoyancy, 146, 147; compressibility, 199; diffusion, 169; dissolved, 99; in ice, 181; outgassing from freezing, 131; in photosynthesis, 80; pressure vs density, 97; supersaturation, 128, 129; thermal expansion, 147. *See also* air; carbon dioxide

giraffes, neckedness, 249

girdling, 132

Glaser, Donald, 131

glass. *See* vitrification

Gould, S. J., 215

Grashof number, 148, 154, 277

grasses, C_3 vs C_4, 78

gravity, 134, 147; droplets on slope, *168*; in flower deployment, 241; tree weight, 230

Greenhill, A. G., 212

ground. *See* soil

growth: leaves, 238–45; trees, 233

guard cells, 81, *83*, 86

Hagen-Poiseuille equation, 109, 110, 112, 113

Hales, Stephen, 94, 107

hardwoods. *See* trees

Hatch-Slack cycle, 76, 78

heat, 3; capacity, 144, 160, 187, 193; conduction, 146; flow, *15*, 140, 176; from decomposing leaves, 162; of fusion, 187, 188, 189; solar, 195; thermal breathing, 156; thermal conductivity, 146, 149, *151*, 193; thermal expansion coefficient, 147, 148, 278; thermodynamic significance, 13; of vaporization, 97, 144, 188. *See also* cooling

Henry's law, 127, 128

herbivory, 4, 177, 179, 245; chewing cellulosics, 249; deterrence, 247–51; energy yield, 249; minimizing payoff, 259–61

Holmes, O.W., 257

Hooke, Robert, 148

humans: auto bodies, 244; beams used by, 202; cake frostings, 193; candy thermometers, 187; circulation, 51, 61, 65, 86, 121; composite material use, 198; cryopreservation, 194, 196; deployable structures, 238; engine designs, 269; erectable penis of, 200; flash-freezing food, 190; food freezer fix, 187; high altitudes and, 53, 54; horse-based cities, 260; household levers, 215; ice skating, 191; kidneys, 86; origami, 244; power output, 12,

72; pressurized cooling systems, 191; ridge-and-valley roofs, 208; structural materials, 173, 198; sweating, 144, 145; teapot spouts, 173; thermal inertia, 140; water-repellent fabrics, 164, 169; water use, 71, 74; woven materials, 198. *See also* animals

humidity, 70, 72, 74, 79, 86, 134, 145, 177

hydrophobicity, 91, 163–70; capillary depression, *163*; contact angle, 164, *165*; demonstrating, 164; droplets on leaves, 167; droplets on slope, 168; insect feet, 166; superhydrophobicity, 165–70, *167*, 178

hydroskeletons, 200–202

ice, 180–96; air in, 196; bubbles in, *183*; coating leaves, 29, 172; crystals, 188, 189, 190; density, 181; extracellular, 187–91, 195; freezing vegetables, 180; frost formation, 26; gases in, 181, 182; ice storms, 212; icicles, 174; mechanical properties, 173; melting, *182*; within organisms, 181, 187–91, 195; skating, 191; solute excluding, 196; thermal conductivity, 183; vitrification, 192–94. *See also* freezing

infrared photography, 22, 23

insects: dispersal, 246; freezing tolerance, 194; as herbivores, 246, 251; at high altitudes, 55; leaf-cutting ants, 252; mandibles, 251; plastrons, 169; superhydrophobic feet, 166; tracheal diffusion, 52; visual spectrum, 21

Kelvin scale, 24

Kolmogorov scale, 60, 62

lakes, 147

Lambert's law, 33

Laplace's law, 118–23, *119*; and cell wall pores, 125; cylinders, 121; deflating balloon, *121*; parallel balloons, *117*;

Raup, David, 239

respiration, 33, 89

Reynolds number, 110, 154, 217, 263, 278

rhododendron curling, 27, 28

roofs, 208–10, 208

roots: bamboo, 235; depth, 134; extracting water, 134–38; root plate, 135, 229; scaling of, 265; soil attachment, 228–36; striker, 230, 231, 234; taproots, 232, 265

Sachs, J. von, 107, 111

safety factor, 214

salt, 139

sap: ascent, 91–115, 122; composition, 102; flow, 94, 107–15; inter-conduit movement, 132; interface in leaves, 125; pressure to drive, 111; viscosity, 109. *See also* water; xylem

scaling. *See* size and scale; piscivory

Scholander bomb, 103–6, 104

Scholander, P. F., 103, 196

science, 5–6; biology, 6–11; biomimetics, 274; forces channeling, 227; physical, 11–16; pure vs applied, 273–77; reductionism, 5–6; as social insurance, 274–77

seawater density, 195

shear stress units, 278

size and scale, 263–67; beams, 204–7; bubbles, 127; cell size, 51, 67, 202; cell wall pores, 125; diffusion utility, 49, 69, 272; folivorous animals, 260; Kolmogorov scale, 64; laminar vs turbulent flow, 59; Laplace's law, 119–22, 201; leaves, 156, 266–67; minimum bubble size, 130; minimum ice crystal size, 189; organism range, 263; surface tension effectiveness, 123, 124; surface-to-volume, 263–66; time scale, 273; tree height, 10, 38–40, 107, 138, 233, 264; water droplets, 171. *See also* dimensionless ratios

sky, radiation to, 25, 29, 78, 145

softwoods. *See* trees

soil: erosion, 170; extracting water from, 134–38, 137; as material, 216; matric pressure, 136, 138; particle surface area, 136; resisting compression, 232; resisting tension, 228, 232, 234; tree uprooting, 228–36

solutions: crystallization, 181, 186; distillation, 186; ionization in, 185; Raoult's law, 185; of sugars, 193. *See also* colligative properties

spores on leaves, 174–77

Stefan-Boltzmann equation, 25, 143, 144, 278

stiffness, 205, 207, 211, 212, 277

Stokes' law, 193

stomata, 52, 82; as chimneys, 66; control, 79–90, 85, 87, 88–90; density of, 81; diffusion through, 52; endogenous rhythms, 89; flow through, 156; humidity near, 177; inputs, 88–90; size, 125; structure, 51, 81, 83, 86

storms. *See* wind

strength: compressive, 210; crushing, 211; Euler buckling, 211; tensile, 210

stress, shear, 57

stress concentration, 254, 256

striker roots, 230, 231, 234

structural materials, 197–202; brittle cracking, 258; cellulose, 197; composites, 198; home-made composite, 199; Mohs' hardness, 250; natural fibers, 126, 197; proteins, 197; toughness, 252; water as, 197; wood, 197

structures: deployable, 209, 238; fanfold stiffening, 209; functions of, 267–69; hydroskeleton model, 200; safety factor, 214; self-assembling, 239. *See also* beams; columns; support

succulents, 160

suction. *See* pressure, negative

sugars, 193, 198

sunflecks, 35

sunlight. *See* light

supercooling, 193–95

superhydrophobicity, 165–70

supersaturation, 128, 129

support, 197–213; banyan trunks, 236; beams, 202–10; columns, 210–13; leaves, 199–210

surface tension, 97, 122–38, 167, 278; area minimizing, 122; capillary rise, *92*, 122, 163, 164; dimensions, 189; Laplace's law, 118–23, *119*; leaf hydrophobicity, 163–70; and microtexture, 165; pressure from, 124; size and scale, 123, 124; water vs air, 123. *See also* bubbles; capillarity

surfaces, 162–79; area minimization, 122; droplets rolling on, 167; energy to make, 189; hairy, 176–78; hydrophilic, 91, 99, 122, 128, 136; hydrophobic, 91, 128, 130; microtextured, 165; relative to volumes, 263–66; soil particle area, 136. *See also* curved surfaces; surface tension

symmorphosis, 257

taproots, *231*

technology. *See* measurement; humans

teeth, 250, *251*

temperature, 277, 278; animal bodies, 161; boiling point, 185; clear sky, 25; freezing point, 185, 186, 188; global average, 181, 195; gradients, 148–52; Kelvin scale, 24; leaves, 22, 27, *28*, 140, 142, 143, 155, 161, 191, 267; liquid crystal sheet, 148, 149, 151; minimum survivable, 194; photosynthetic optimum, 140; stomatal stimulus, 89; supercooling, 184; vs wavelength, 24

tensile strength: solids, 101, 202; water, 97–102

tension, 278; balloon walls, 118; in cantilever, 203, 204; vs compression, 198;

vs pressure, 116–24; withstanding, 204. *See also* Laplace's law, surface tension

thermal. *See* heat

thermodynamics: demonstrations, 12, 13–15; of diffusion, 46; efficiency limits, 73; first law, 11, 13, 147; heat, role of, 13; light energy, 21; Maxwell's demon, 13; osmosis, 85; second law, 12–16, 140

Thompson, D'Arcy, 212

torque. *See* turning moment

toughness, 252, 253

transpiration. *See* evaporation

trees: buttresses, 233; as C_3 users, 77; as cantilever beams, 212–13, 215, 227–36, 265; as columns, 210–13; crown drag, 213; defoliation tolerance, 246; footplates, 135; girdling, 132; height, 10, 38–40, 107, 138, 210–13, 233, 64, 272; in high winds, 214–16, 218; layers of leaves, 35–38; as levers, 216; life history, 214; mechanical failure, 214–16; midday girth shrinkage, 93, 96; multi-trunked, 236, 237; negative pressures in, 94–107; of rain forests, 232, 234; repairing embolisms, 133; role of weight, 230; root depth, 134; safety factor, 214; swaying, 231, 232, 237; trunk failure, 210; turning by wind, *216*, 227–28, 230; uprooting, 210, 228–36; water ascent, 71, 91–115. *See also* forests

trusses. *See* beams

turning moment, 227–28, 230

uprooting, 210, 228–36

vapor pressure, 84, 185

vaporization. *See* evaporation

vectors, 228

veins, 202, 207

velocity gradients, 61; in convection,

148; on leaf surfaces, 62, 174, 175; thickness, 62, 63, 278

viscosity, 57–62, 278; air, 148; corn syrup, 193; in Hagen-Poiseuille equation, 109; in Reynolds number, 110; sap, 109, 111

vitreous humor. *See* bull in china shop

vitrification, 192–94, 195

water: air interface, 116; ascent in trees, 71, 91–115; atmospheric, 42; boiling, *130*; bulk modulus, 199; C_3 vs C_4 plant use, 77; CAM plant use, 78; capillarity, 91–93, *92*, 122, 163, 164; density, 144, 172, 195, 217; diffusion, 47, 70–72, 80; droplets, 124; forest's use, 145; freezable fraction, 190; gas solubility, 182; human use, 71, 74; hydroskeletons, 200–202; internal cohesion, 122; leaf drip tips, 170–74; liquid phase, 181; loss from leaf, 70–90, *80*; loss reduction by hairiness, 177; loss vs CO_2 loss, 72–74; osmotic pressures, 138; for photosynthesis, 70; seawater, 139, 195; in soil, 134–38, *137*; stomatal stimulus, 89; as structural material, 197, 199; supercooling, 195; surface tension, 97; tensile strength, 97–102, *100*, 128; thermal properties, 97, 144, 147, 160, 187, 188; use efficiency, 74–81, 90, 278; vapor in air, 70; vitrification, 192, 195. *See also* flow; fluids; rain

wavelength, 19, 24

weaving, 198

Wien's law, 25, 143

Wigner, Eugene, 5

wind: forced convection, 152–56; gusts, 237; imperceptible, 155; irregularity of, 175; leaf drag, 217–27; on leaves, 65; leaves reconfiguring, 220, 221, 223, 226; leaves in storms, 224; maximum tolerable, 225; minimal perceptible, 154; on spores on leaves, 174; storm intensity, 214–37; swaying trees, 231, 232; tearing leaves, 257; vs tree height, 265; on trees, 212–13, 216; turbulence, 221. *See also* convection; flow

wind tunnels, 218

wood: as food, 259; compressive strength, 210; density, 211, 229; as structural material, 197; tensile strength, 202, 210

work, 11

xylem: cavitation, 112; conduits, 102–3, 113, 114, 131, 133; diameters, 93, 108; embolisms, 106, 112; flow in, 107–15, 138; freezing of, 106; hydrophilicity, 102; keeping air out, 116–26, 131; multifunctionality, 268; pressures within, 103–6; reestablishing continuity, 99; width vs water use, 111, 113

Young's modulus, 205, 207, 211, 212, 277

Young-Laplace equation. *See* Laplace's law